Hands-on Machine Learning with Python

Implement Neural Network Solutions with Scikit-learn and PyTorch

Ashwin Pajankar
Aditya Joshi

Apress®

Hands-on Machine Learning with Python: Implement Neural Network Solutions with Scikit-learn and PyTorch

Ashwin Pajankar
Nashik, Maharashtra, India

Aditya Joshi
Haldwani, Uttarakhand, India

ISBN-13 (pbk): 978-1-4842-7920-5
https://doi.org/10.1007/978-1-4842-7921-2

ISBN-13 (electronic): 978-1-4842-7921-2

Managing Director, Apress Media LLC: Welmoed Spahr
Acquisitions Editor: Celestin Suresh John
Development Editor: James Markham
Coordinating Editor: Aditee Mirashi

Cover designed by eStudioCalamar

Cover image designed by Freepik (www.freepik.com)

Distributed to the book trade worldwide by Springer Science+Business Media New York, 1 New York Plaza, Suite 4600, New York, NY 10004-1562, USA. Phone 1-800-SPRINGER, fax (201) 348-4505, e-mail orders-ny@ springer-sbm.com, or visit www.springeronline.com. Apress Media, LLC is a California LLC and the sole member (owner) is Springer Science + Business Media Finance Inc (SSBM Finance Inc). SSBM Finance Inc is a **Delaware** corporation.

For information on translations, please e-mail booktranslations@springernature.com; for reprint, paperback, or audio rights, please e-mail bookpermissions@springernature.com.

Apress titles may be purchased in bulk for academic, corporate, or promotional use. eBook versions and licenses are also available for most titles. For more information, reference our Print and eBook Bulk Sales web page at http://www.apress.com/bulk-sales.

Any source code or other supplementary material referenced by the author in this book is available to readers on GitHub via the book's product page, located at www.apress.com/978-1-4842-7920-5. For more detailed information, please visit http://www.apress.com/source-code.

Printed on acid-free paper

This book is dedicated to the memory of our teacher, Prof. Govindarajulu Regeti (July 9, 1945–March 18, 2021)

Popularly known to everyone as RGR, Prof. Govindarajulu obtained his B.Tech. in Electrical and Electronics Engineering from JNTU Kakinada. He also earned his M.Tech. and Ph.D. from IIT Kanpur. Prof. Govindarajulu was an early faculty member of IIIT Hyderabad and played a significant role in making IIIT Hyderabad a top-class institution that it grew to become today. He was by far the most loved and cheered for faculty member of the institute. He was full of energy to teach and full of old-fashioned charm. There is no doubt he cared for every student as an individual, taking care to know about and to guide them. He has taught, guided, and mentored many batches of students at IIIT Hyderabad (including one of the authors of the book, Ashwin Pajankar).

Table of Contents

About the Authors

Ashwin Pajankar is an author, an online instructor, a content creator, and a YouTuber. He has earned a Bachelor of Engineering from SGGSIE&T Nanded and an M.Tech. in Computer Science and Engineering from IIIT Hyderabad. He was introduced to the amazing world of electronics and computer programming at the age of seven. BASIC is the very first programming language he learned. He has a lot of experience in programming with Assembly Language, C, C++, Visual Basic, Java, Shell Scripting, Python, SQL, and JavaScript. He also loves to work with single-board computers and microcontrollers like Raspberry Pi, Banana Pro, Arduino, BBC Microbit, and ESP32.

He is currently focusing on developing his YouTube channel on computer programming, electronics, and microcontrollers.

Aditya Joshi is a machine learning engineer who's worked in data science and ML teams of early to mid-stage startups. He has earned a Bachelor of Engineering from Pune University and an M.S. in Computer Science and Engineering from IIIT Hyderabad. He became interested in machine learning during his masters and got associated with the Search and Information Extraction Lab at IIIT Hyderabad. He loves to teach, and he has been involved in training workshops, meetups, and short courses.

About the Technical Reviewer

Joos Korstanje is a data scientist, with over five years of industry experience in developing machine learning tools, of which a large part is forecasting models. He currently works at Disneyland Paris where he develops machine learning for a variety of tools.

Acknowledgments

I would like to express my gratitude toward Aditya Joshi, my junior from IIIT Hyderabad and now an esteemed colleague who has written the major and the most important section of this book. I also wish to thank my mentors from Apress, Celestin, Aditee, James Markham, and the editorial team. I wish to thank the reviewers who helped me make this book better. I also thank Prof. Govindrajulu's family – Srinivas (son) and Amy (daughter-in-law) – for allowing me to dedicate this book to his memory and sharing his biographical information and his photograph for publication.

—Ashwin Pajankar

My work on this book started with a lot of encouragement and support from my father, Ashok Kumar Joshi, who couldn't live long enough to see it till completion. I am extremely grateful to friends and family – especially my mother, Bhavana Joshi, and many others, whose constant support was the catalyst to help me work on this project. I also want to extend my heartiest thanks to my wife, Neha Pandey, who was supportive and patient enough when I extended my work especially during weekends. I would like to thank Ashwin Pajankar, who's been not just a coauthor but a guide throughout this journey. I'd also like to extend my gratitude to the Innomatics team, Kalpana Katiki Reddy, Vishwanath Nyathani, and Raghuram Aduri, for giving me opportunities to interact with hundreds of students who are learning data science and machine learning. I'd also like to thank Akshaj Verma for his support with code examples in one of the advanced chapters. I also thank the editorial team at Apress, especially Celestin Suresh John, Aditee Mirashi, James Markham, and everyone who was involved in the process.

—Aditya Joshi

Introduction

We have long been planning to collaborate and write a book on machine learning. This field has grown and expanded immensely since we started learning these topics almost a decade ago. We realized that, as lifelong learners ourselves, the initial few steps in any field require a much clearer source that shows a path clearly. This also requires a crisp set of explanation and occasional ideas to expand the learning experience by reading, learning, and utilizing what you have learned. We have used Python for a long duration in our academic life and professional careers in software development, data science, and machine learning. Through this book, we have made a very humble attempt to write a step-by-step guide on the topic of machine learning for absolute beginners. Every chapter of the book has the explanation of the concepts used, code examples, explanation of the code examples, and screenshots of the outputs.

The first chapter covers the setup of the Python environment on different platforms. The second chapter covers NumPy and Ndarrays. The third chapter explores visualization with Matplotlib. The fourth chapter introduces us to the Pandas data science library. All these initial chapters build the programming and basic data crunching foundations that are one of the prerequisites for learning machine learning.

The next section discusses traditional machine learning approaches. In Chapter 5, we start with a bird's-eye view of the field of machine learning followed by the installation of Scikit-learn and a short and quick example of a machine learning solution with Scikit-learn. Chapter 6 elaborates methods to help you understand and transform structural, textual, and image data into the format that's acceptable by machine learning libraries. In Chapter 7, we introduce supervised learning methods, starting with linear regression for regression problems and logistic regression and decision trees for classification problems. In each of the experiments, we also show how to plot visualizations that the algorithm has learned with the use of decision boundary plots. The eighth chapter ponders over further fine-tuning of machine learning models. We explain some ideas for measuring the performance of the models, issues of overfitting and underfitting, and approaches for handling such issues and improving the model performance. The ninth chapter continues the discussion of supervised learning methods especially focusing on naive Bayes and Support Vector Machines. The tenth chapter explains ensemble learning methods, which are the solutions that combine multiple simpler models to produce a performance better than what they might offer individually. In the eleventh chapter, we discuss unsupervised learning methods, specifically focusing on dimensionality reduction, clustering, and frequent pattern mining methods. Each part contains a complete example of implementing the discussed methods using Scikit-learn.

The last section begins with introducing the basic ideas of neural network and deep learning in the twelfth chapter. We introduce a highly popular open source machine learning framework, PyTorch, that will be used in the examples in the subsequent chapters. The thirteenth chapter begins with the explanation of artificial neural networks and thoroughly discusses the theoretical foundations of feedforward and backpropagation, followed by a short discussion on loss functions and an example of a simple neural network. In the second half, we explain how to create a multilayer neural network that is capable of identifying handwritten digits. In the fourteenth chapter, we discuss convolutional neural networks and work through an example for image classification. The fifteenth chapter discusses recurrent neural networks and walks you through a sequence modeling problem. In the final, sixteenth chapter, we discuss strategies for planning, managing, and engineering machine learning and data science projects. We also discuss a short end-to-end example of sentiment analysis using deep learning.

If you are new to the subject, we highly encourage you to follow the chapters sequentially as the ideas build upon each other. Follow through all the code sections, and feel free to modify and tweak the code structure, datasets, and hyperparameters. If you already know some of the topics, feel free to skip to the topics of your interest and examine the relevant sections thoroughly. We wish you the best for your learning experience.

SECTION 1

Python for Machine Learning

CHAPTER 1

Getting Started with Python 3 and Jupyter Notebook

I hope that all of you have read the introduction and the table of contents. This is very important because if you are a complete beginner, please do not skip this chapter. The entire field of machine learning and artificial intelligence requires solid knowledge of the tools and frameworks used in the area. This chapter serves as the foundational chapter for Python programming for machine learning that we will cover in this book. It introduces the novice readers to the Python Programming Language, Scientific Python Ecosystem, and Jupyter Notebook for Python programming.

The following is the list of topics that we will learn in this chapter:

- Python 3 Programming Language

- Installing Python

- Python Modes

- Pip3 Utility

- Scientific Python Ecosystem

- Python Implementations and Distributions

After studying this chapter, we will be comfortable with the installation, running programs, and Jupyter notebook on Windows and Debian Linux.

© Ashwin Pajankar and Aditya Joshi 2022
A. Pajankar and A. Joshi, *Hands-on Machine Learning with Python*, https://doi.org/10.1007/978-1-4842-7921-2_1

Python 3 Programming Language

Python 3 is a modern programming language. It has features of object-oriented and procedural programming. It runs on a variety of platforms. The most readily available platforms for common readers are macOS, Windows, and various Linux distributions. Python runs on all of them. And the major advantage is that the code written on one platform runs on the other platform without any major changes to the code (except for the platform-specific code). You can read more about Python at `www.python.org/`.

History of Python Programming Language

The ABC language is the predecessor of Python Programming Language. The ABC language was inspired by ALGOL 68 and SETL programming languages. The Python Programming Language was created by Guido Van Rossum as a side project during Christmas vacations in the late 1980s. He was working at the National Research Institute for Mathematics and Computer Science (Centrum Wiskunde & Informatica). Van Rossum was a graduate of the University of Amsterdam in computer science. He has worked for Google and Dropbox. Currently, he works for Microsoft.

Python has two major and incompatible versions: Python 2 and Python 3. Python 2 is now not under active development and maintenance. The entire (programming) world is gradually switching to Python 3 from Python 2. For all the demonstrations in this book, we will use Python 3. Whenever we use the word **Python**, it will mean **Python 3** from now onward.

Philosophy of Python Programming Language

The philosophy of Python Programming Language is known as **The Zen of Python**, and it can be accessed at `www.python.org/dev/peps/pep-0020/`. The following are the points from that PEP (Python Enhancement Proposal). A few are funny.

1. Beautiful is better than ugly.

2. Explicit is better than implicit.

3. Simple is better than complex.

4. Complex is better than complicated.

5. Flat is better than nested.

6.   Sparse is better than dense.

7.   Readability counts.

8.   Special cases aren't special enough to break the rules.

9.   Although practicality beats purity.

10.   Errors should never pass silently.

11.   Unless explicitly silenced.

12.   In the face of ambiguity, refuse the temptation to guess.

13.   There should be one – and preferably only one – obvious way
       to do it.

14.   Although that way may not be obvious at first unless you're Dutch.

15.   Now is better than never.

16.   Although never is often better than *right* now.

17.   If the implementation is hard to explain, it's a bad idea.

18.   If the implementation is easy to explain, it may be a good idea.

19.   Namespaces are one honking great idea – let's do more of those!

These are general philosophical guidelines that influenced the development of the Python Programming Language over decades and continue to do so.

Where Python Is Used

Python is used in a variety of applications. A few are

1)   Education

2)   Automation

3)   Scientific Computing

4)   Computer Vision

5)   Animation

6)   IoT

7) Web Development

8) Desktop and Mobile Applications

9) Administration

We can read all the applications in detail at `www.python.org/about/apps/`. Many organizations have used Python to create applications. We can read all these success stories at `www.python.org/success-stories/`. Now, let's begin with our very own one.

Installing Python

We will learn how to install Python 3 on Windows in detail. Visit `www.python.org` and hover the mouse pointer over the **Downloads** options. It will open the downloads menu. It will show the appropriate option depending on your OS. In our case, it will show option for downloading on Windows. Download the file. It is an executable installation file. In my case, it downloads a 64-bit version of the installable file. If you are using other architecture (e.g., 32 bit), then it will download the appropriate file accordingly. Figure 1-1 shows the Python 3 download for Windows.

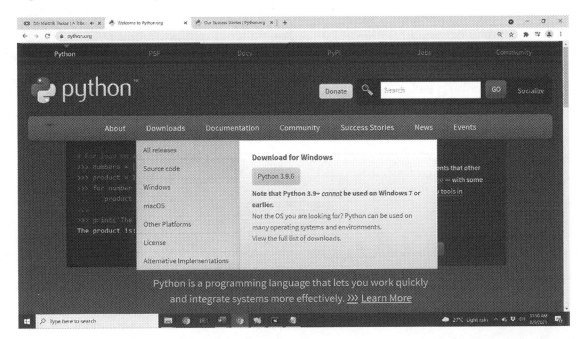

Figure 1-1. *Python 3 download for Windows*

Once downloaded, open the file. It will show a window shown in Figure 1-2. Do not forget to check all the checkboxes so that the Python's installation folder can be added to the **PATH** variable in the Windows environment. It enables us to launch Python from the command prompt.

Figure 1-2. *Python 3 installation for Windows (check all the checkboxes)*

Click on the Install Now option (it requires the administrators' privileges). After installation finishes successfully, it shows a message shown in Figure 1-3.

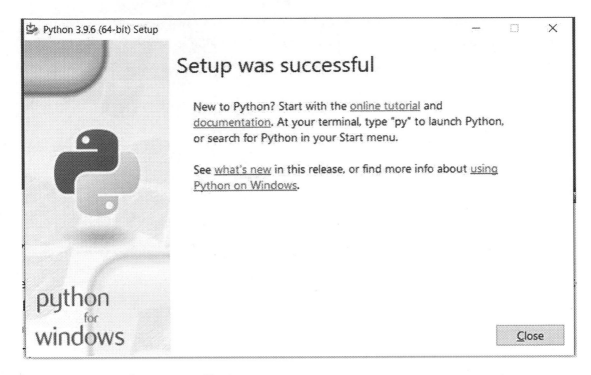

Figure 1-3. *Python 3 installation success message*

Close the window now, and we are ready for our journey.

Python on Linux Distributions

Python 2 and Python 3 come pre-installed on all the major Linux distributions. We will see that later in this chapter.

Python on macOS

We can get detailed instructions of the installation on macOS at `https://docs.python.org/3/using/mac.html`.

Python Modes

Let us study various Python modes. We will also write our first Python program.

Interactive Mode

Python provides us interactive mode. We can invoke Python interactive mode by opening the IDLE (Integrated Development and Learning Editor) program that comes with the Windows installation. Just type the word IDLE in the Windows search bar and click the IDLE icon that appears as shown in Figure 1-4.

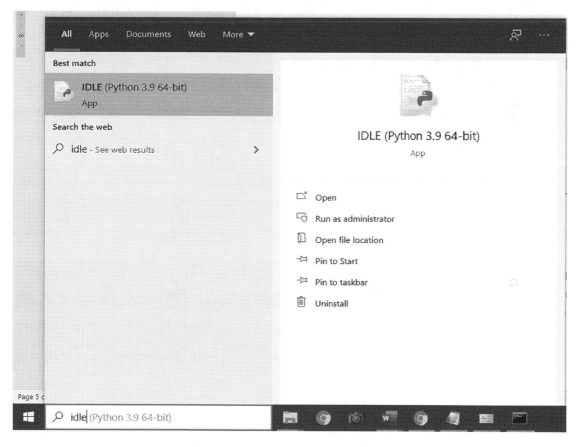

Figure 1-4. *Launching IDLE on Windows*

It will show a window as shown in Figure 1-5.

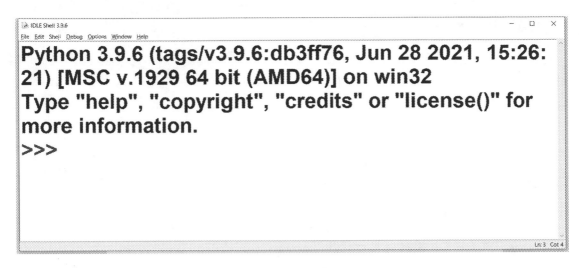

Figure 1-5. *IDLE interactive prompt on Windows*

We can type in the following code on that:

```
print("Hello, World!")
```

Then hit the enter key to run that. It will show the output as shown in Figure 1-6.

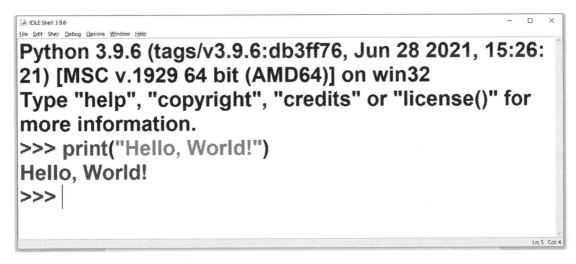

Figure 1-6. *Simple code execution*

Now we know that IDLE can execute the code line by line or block by block. It is very convenient to run small independent snippets of code here in the interactive prompt.

We can also invoke the interpreter mode without IDLE from the command line by typing in the command in Windows command prompt. The command is `python`. Run the command, and it will launch the interpreter in the interactive mode shown in Figure 1-7.

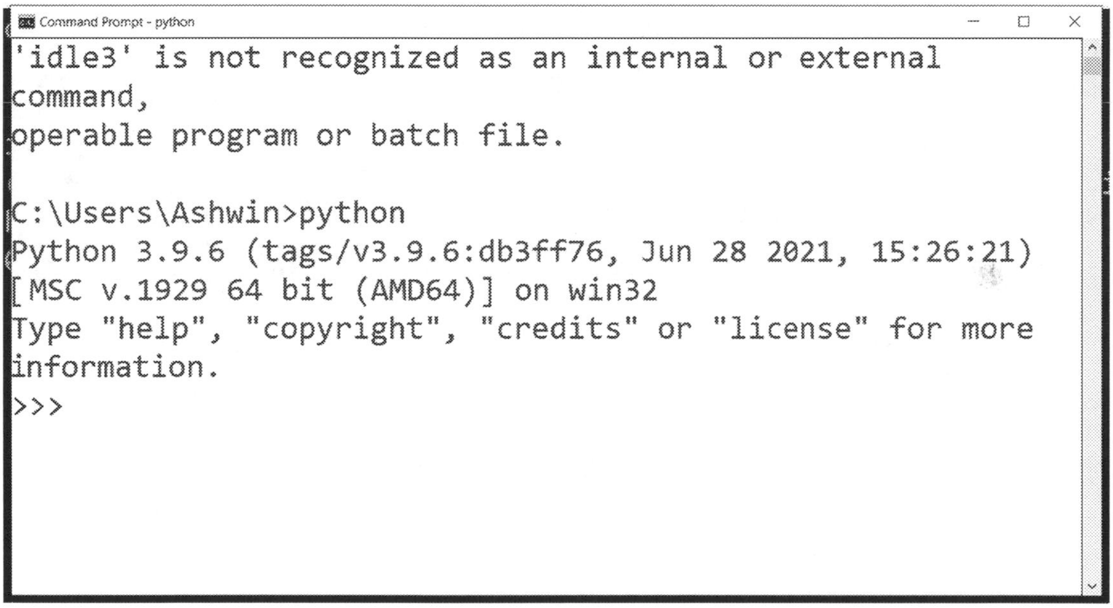

```
Command Prompt - python                                    —   □   ×

'idle3' is not recognized as an internal or external
command,
operable program or batch file.

C:\Users\Ashwin>python
Python 3.9.6 (tags/v3.9.6:db3ff76, Jun 28 2021, 15:26:21)
[MSC v.1929 64 bit (AMD64)] on win32
Type "help", "copyright", "credits" or "license" for more
information.
>>>
```

Figure 1-7. *IDLE on Windows command prompt*

I use a flavor of Debian distribution (Raspberry Pi OS) on a Raspberry Pi 4 with 8GB of RAM. IDLE does not come pre-installed, but Python 2 and Python 3 are there in the OS. Run the following command on the command prompt (terminal or using an SSH client) to install IDLE for Python 3:

```
sudo pip3 install idle
```

It will install IDLE for Python 3. We will discuss the **pip** utility later in this chapter. Also, without IDLE too, we can invoke the Python 3 interactive mode by typing in the command `python3` on the command prompt on Linux. Note that as the Linux distribution comes with both Python 2 and Python 3, the command `python` will invoke Python 2 on the command prompt. For the Python 3 interpreter, the command is `python3` on Linux. Figure 1-8 shows a Python 3 session in progress in the SSH terminal accessing the Linux command prompt remotely.

11

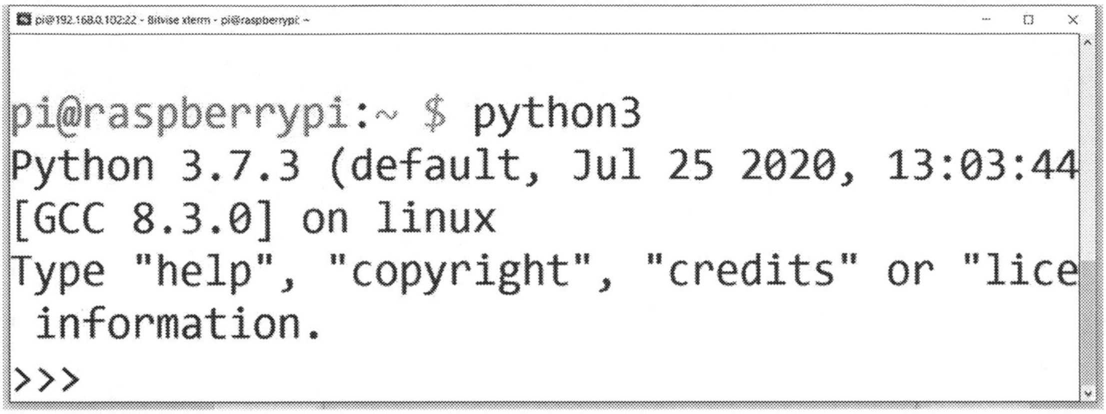

```
pi@raspberrypi:~ $ python3
Python 3.7.3 (default, Jul 25 2020, 13:03:44
[GCC 8.3.0] on linux
Type "help", "copyright", "credits" or "lice
 information.
>>>
```

Figure 1-8. Python 3 interpreter on the Linux command prompt (remote SSH access)

Now, we can access the IDLE for Python 3 by typing in the command `idle` on the command prompt invoked in the Linux desktop. If we run this command remotely in the SSH terminal, it returns an error as the remote SSH lacks GUI features. We can run this command on the terminal only invoked from the desktop environment of Linux to invoke the IDLE. It could be directly done or even done in a remote desktop environment like VNC as shown in Figure 1-9.

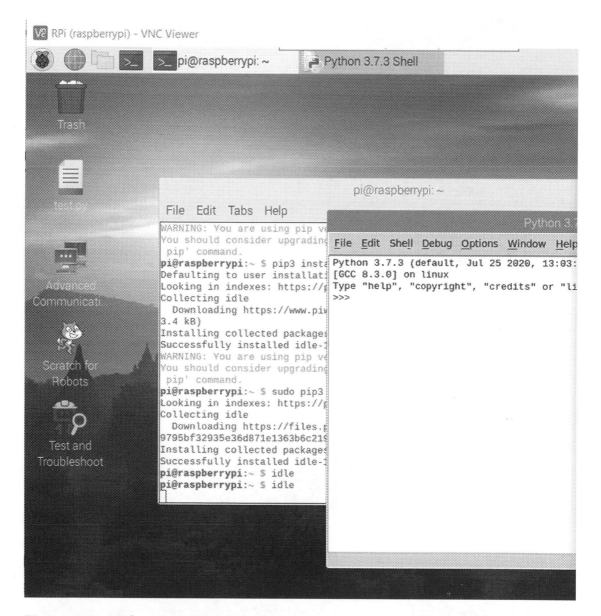

Figure 1-9. *Python 3 IDLE invoked on the Linux command prompt (remote Desktop access with VNC)*

We can also launch the IDLE from the Raspberry Pi OS menu as shown in Figure 1-10.

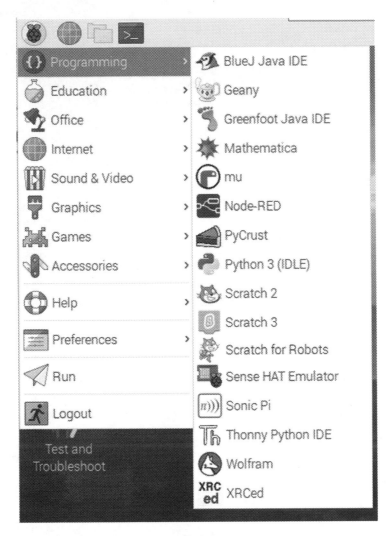

Figure 1-10. *IDLE under the Linux menu*

Whatever flavor of Linux you are using, you can definitely find the IDLE in the menu and programs once installed. If you do not find it, launch it from the command prompt in the desktop as discussed earlier.

Python 3 also has a more interactive command line environment known as IPython. We can install it by running the following command in the command prompt of Linux and Windows:

```
pip3 install ipython
```

It will install IPython for Python 3 on your OS. We can invoke IPython by typing the command `ipython` in the command prompts of Windows and Linux. In place of IDLE interactive mode, we can also use this. Figure 1-11 shows an IPython session in progress.

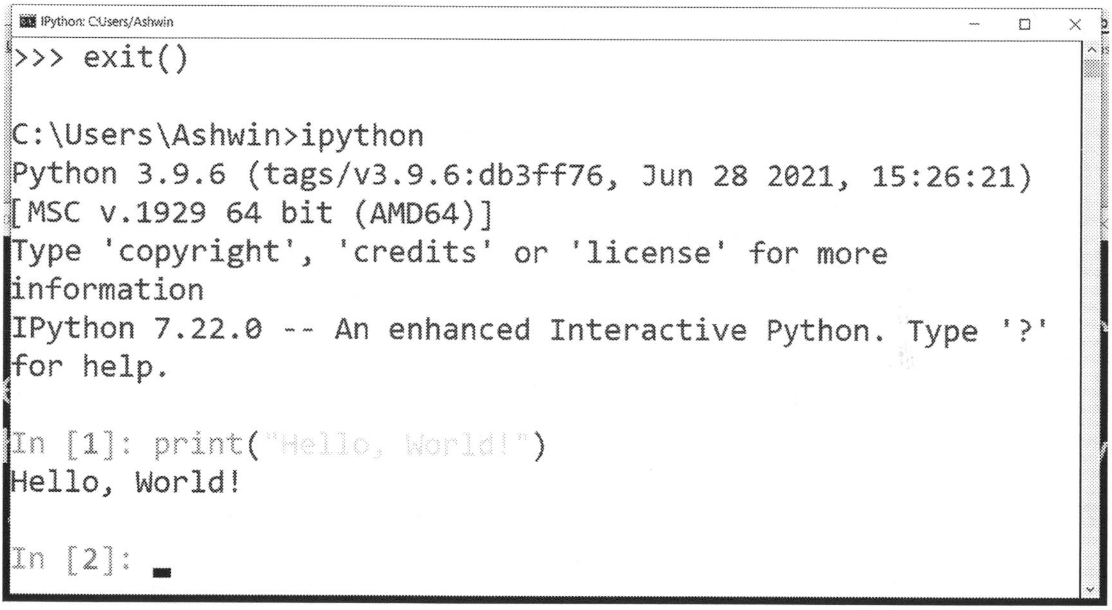

```
IPython: C:Users/Ashwin                                    —    □    ×
>>> exit()

C:\Users\Ashwin>ipython
Python 3.9.6 (tags/v3.9.6:db3ff76, Jun 28 2021, 15:26:21)
[MSC v.1929 64 bit (AMD64)]
Type 'copyright', 'credits' or 'license' for more
information
IPython 7.22.0 -- An enhanced Interactive Python. Type '?'
for help.

In [1]: print("Hello, World!")
Hello, World!

In [2]: _
```

Figure 1-11. *An IPython session under progress in a Windows command prompt*

Finally, the following command terminates the Python interpreter, IDLE interactive mode, and IPython sessions in all the platforms (Windows, macOS, and Linux):

```
exit()
```

This is all about the interactive mode and Python interpreter for now.

Script Mode

Interactive mode of the Python interpreter is easy to get started with and is very good for prototyping and so on. However, for bigger programs, the best way is to use IDLE. Actually, we can use any editor of our choice to write Python programs. However, plaintext editors like Notepad or gedit cannot run the programs. So we use IDEs like IDLE. IDLE is easier to use. Just click the **New File** option in the **File** menu in the menu bar at the top of the IDLE interpreter window, and it creates a new blank file. We have to save it. IDLE automatically adds ***.py** extension to the saved files. Then type in the following line of code and save again:

```
print("Hello, World!")
```

Then go to the **Run** menu in the menu bar and then click the **Run Module** option. We can also directly run using the shortcut **F5** key on the keyboard. Figure 1-12 shows an IDLE window and its output.

Figure 1-12. *A Python program in IDLE and the output of the program*

We can also launch the program from the command prompts of the operating systems. In the command prompt, just open the directory (with the command **cd**) that has the Python program saved and run the following command:

```
python3 prog01.py
```

In Windows, we need to run the following command:

```
python prog01.py
```

The Python interpreter will run the program, show the output, and return the control to the command prompt.

In Linux, we can also directly run the Python code file in the command prompt without invoking the interpreter directly. For that, we need to add the following line to the code in the beginning of the file:

```
#!/usr/bin/python3
```

So the code file finally looks like the following:

```
#!/usr/bin/python3
print("Hello, World!")
```

In the command prompt, open the directory (again with the command **cd**) and run the following command to change the mode of file:

```
chmod 755 prog01.py
```

It will make it executable. We can now directly invoke the program file with the following command:

```
./prog01.py
```

The Linux Shell will use the Python interpreter mentioned in the first line of the code file to execute it.

To find out the location of the executable file for the Python interpreter in your Linux distribution, run the following command on the command prompt:

```
which python3
```

In Windows, the same can be done with

```
where python
```

It will return the location of the executable file for the Python interpreter in Windows.

Pip3 Utility

We have installed a couple of things with the pip utility earlier. In this section, we will learn the utility in detail.

Pip stands for **pip installs packages** or **pip installs Python**. It is a recursive acronym. It means that the expansion of the terms includes itself. Pip is a package manager utility for Python and related software. It is a command line utility. Pip3 is the Python 3 version of pip. We can use it to install packages for Python 3. In the entire book, we will use it throughout to install libraries. Let us see the usage.

We can use it to see the list of installed packages as follows:

```
pip3 list
```

It will show the list of all the packages installed. We can install a new package (in this case, Jupyter) as follows:

```
pip3 install jupyter
```

In Linux, we may have to use it with sudo utility as follows:

```
sudo pip3 install jupyter
```

We can also uninstall packages with the following command:

```
pip3 uninstall jupyter
```

Scientific Python Ecosystem

We can use Python for numerical and scientific programming. We have to use the Scientific Python Ecosystem. It has the following core components:

1) Python

2) Jupyter Notebook

3) NumPy

4) SciPy

5) Matplotlib

6) SymPy

We will be using most of the components except SymPy.

We have already seen how to install Jupyter in the last section. Jupyter is a web browser–based interpreter for Python. Install it if you have not installed it yet.

We can launch Jupyter with the following command in the command prompts of Windows and Linux OS:

```
jupyter notebook
```

It will launch the Jupyter Notebook server process and open a web browser–based interface automatically. As an exercise to this chapter, explore Jupyter Notebook and run small code snippets in that.

Python Implementations and Distributions

We are comfortable with the basics of Python and have set up our computers for Python programming. I believe that now is the time we should explore the concepts, **Python Implementations** and **Python Distributions**. These concepts are essential to know for beginners.

Earlier, I have mentioned that Python is a programming language. However, it is more than that. We all know about **C** programming language. **C** is more of a programming standard. The standard is decided by ANSI, and various organizations write their own programs and tools (known as compilers) that compile C programs as per the requirements. I have written C programs with various compilers (and associated IDEs) like Turbo C++, Microsoft Visual C++, GCC, LLVM, and MinGW-w64 on Windows.

I believe that Python is also evolving in a similar direction as many organizations have written their own interpreters that interpret and execute Python code. These interpreters are also known as **Implementations** of Python. Earlier in the chapter, we learned how to download and install Python interpreter (and IDLE) from `www.python.org`. This interpreter is also known as **CPython Implementation**. It is written in C and Python. It is the reference implementation of Python. There are several other implementations. The following is a list (non-exhaustive) of other implementations of Python:

- IronPython (Python running on .NET)

- Jython (Python running on the Java Virtual Machine)

- PyPy (a fast Python implementation with a JIT compiler)

- Stackless Python (branch of CPython supporting microthreads)

- MicroPython (Python running on micro controllers)

We can read more about these at the following URLs:

```
www.python.org/download/alternatives/
https://wiki.python.org/moin/PythonImplementations
```

Now that we are aware that Python has many implementations, let's understand the meaning of **Distribution**. A Python distribution is a Python implementation (interpreter) bundled with useful libraries and tools. For example, the CPython implementation and distribution comes with IDLE and pip. So we have a set of an interpreter, an IDE, and a package manager to get started with the software development. The following URL has many Python distributions listed:

`https://wiki.python.org/moin/PythonDistributions`

These lists of Python implementations and distributions of Python is obviously not exhaustive. Even we can write our own interpreter of Python (it is not an easy task but it can be done). We can package it with a few useful tools and call it our own distribution. I have learned so much about the terms **implementation** and **distribution** from the following stackoverflow.com discussion:

`https://stackoverflow.com/questions/27450172/python-implementation-vs-python-distribution-vs-python-itself`

I encourage all the readers to visit the URL and read the discussion.

Anaconda

Since the book is about the topics in the area of **machine learning** that itself is a part of the broader area of **scientific computing**, this chapter won't be complete unless we discuss the **Anaconda** Python distribution. Anaconda distribution is tailored for the needs of the scientific community, and it comes with a large set of libraries, so beginners do not have to install anything to get started with scientific computing. It also comes with pip as well as another powerful package manager, **conda**. We can download the individual edition at `www.anaconda.com/products/individual`.

It will download an installable file appropriate for your OS (just like Python homepage, it automatically detects your OS and presents you the most suitable file for installation). Go ahead and install it. It is as simple as we installed the reference implementation, CPython. Just make sure that you check the option of adding Anaconda to the **PATH** variable when asked during the installation. Once installed, you can start programming with a lot of tools and IDEs like PyCharm, Spyder, and Jupyter Notebook.

You can open the **Anaconda Navigator** by searching for that term in the search box of Windows. Anaconda Navigator is the unified interface of the Anaconda distribution, and it shows all the tools and utilities in a single window from where we can manage and use them.

However, at this stage, I would like to add the cautionary note. In case you have not yet installed CPython or Anaconda or any other distribution of Python, make sure that you install only a single (and not multiple) distribution of Python as it may be a bit confusing for a beginner to manage multiple Python distributions and environments on a single computer. So for practicing the code examples in the book, install either CPython or Anaconda, but not both, not at least at the same time.

Also, there is a light version of this Anaconda, and it is known as **Miniconda**. It comes with Python interpreter and the conda package manager. It also has a few other useful packages like pip, zlib, and a few others. We can read more about it at `https://docs.conda.io/en/latest/miniconda.html`.

If you wish to read more about the conda package manager, then read the documentation at `https://docs.conda.io/projects/conda/en/latest/commands.html`. It has a comparison table of commands for conda and pip for managing Python packages.

Also, in case you want a detailed step-by-step tutorial and exploration guide, you can watch the video created by me on my YouTube channel at `www.youtube.com/watch?v=Cs4LawOFQ2E`.

Summary

In this chapter, we familiarized ourselves with Python, IDLE, and Anaconda. We learned how to use Python interpreter and how to write Python scripts. We also learned how to execute Python programs. In the next chapter, we will continue our journey with NumPy.

## CHAPTER 2

Getting Started with NumPy

We learned the basics of Python Programming Language in the previous chapter. This chapter is fully focused on learning the basics of NumPy library. We will have a lot of hands-on programming in this chapter. While the programming is not very difficult when it comes to NumPy and Python, the concepts are worth learning. I recommend all readers to spend some time to comprehend the ideas presented in this chapter.

One of the major prerequisites of this chapter is that readers should have explored Jupyter Notebook for Python programming. If you have not already (as I prescribed toward the end of the previous chapter), I recommend you to learn to use it effectively. The following is the list of web pages teaching how to use Jupyter Notebook effectively:

```
www.dataquest.io/blog/jupyter-notebook-tutorial/
https://jupyter.org/documentation
https://realpython.com/jupyter-notebook-introduction/
www.tutorialspoint.com/jupyter/jupyter_notebook_markdown_cells.htm
www.datacamp.com/community/tutorials/tutorial-jupyter-notebook
```

This entire chapter solely focuses on NumPy and its functionalities. The chapter covers the following topics:

- Getting started with NumPy

- Multidimensional Ndarrays

- Indexing of Ndarrays

- Ndarray properties

- NumPy constants

© Ashwin Pajankar and Aditya Joshi 2022
A. Pajankar and A. Joshi, *Hands-on Machine Learning with Python*, https://doi.org/10.1007/978-1-4842-7921-2_2

After studying this chapter, we will be comfortable with the basic aspects of programming with NumPy.

Getting Started with NumPy

I hope that you are comfortable with Jupyter enough to start writing small snippets. Create a new Jupyter Notebook for this chapter, and we will always write our code for each chapter in separate Jupyter Notebooks. This will keep the code organized for reference.

We can run an OS command in the Jupyter Notebook by prefixing it with exclamation mark as follows:

```
!pip3 install numpy
```

We know that Python 3 and pip3 are accessible in Linux by default, and we added Python's installation directory in the system environment variable PATH in Windows OS while installing. That is why the command we just executed should run without any errors and install NumPy library to your computer.

Note   In case the output of the execution of the command shows a warning saying that the pip has a new version, you may wish to upgrade your pip utility with the following command:

```
!pip3 install --upgrade --user pip
```

Remember that this will not have any impact on the libraries we install or the code examples we demonstrate. Libraries are fetched from the repository at https://pypi.org/ (also known as Python Package Index), and any version of pip utility can install the latest version of the libraries.

Once we are done installing the NumPy (and upgrading the pip if you choose to do so), we are ready to start programming with NumPy. But wait! What is NumPy and why are we learning it? How is it related to machine learning? All these questions must be bothering you since you started reading this chapter. Let me answer them.

NumPy is the fundamental library for the numerical computation. It is an integral part of the Scientific Python Ecosystem. If you wish to learn any other library of the ecosystem, I (or any seasoned professional for that matter) will recommend you learn

NumPy first. NumPy is important because it is used to store the data. It has a basic yet very versatile data structure known as Ndarray. It means **N Dimensional Array**. Python has many array-like data structures (e.g., list). But Ndarray is the most versatile and the most preferred structure to store scientific and numerical data.

Many libraries have their own data structures, and most of them use Ndarrays as their base. And Ndarrays are compatible with many data structures and routine just like the lists. We will see the examples of these in the next chapter. But for now, let's focus on Ndarrays.

Let us create a simple Ndarray as follows:

```
import numpy as np
lst1 = [1, 2, 3]
arr1 = np.array(lst1)
```

Here, we are importing NumPy as an alias. Then, we are creating a list and passing it as an argument to the function `array()`. Let's see the data types of all the variables used:

```
print(type(lst1))
print(type(arr1))
```

The output is as follows:

```
<class 'list'>
<class 'numpy.ndarray'>
```

Let's see the contents of the Ndarray as follows:

```
arr1
```

The output is as follows:

```
array([1, 2, 3])
```

We can write it in a single line as follows:

```
arr1 = np.array([1, 2, 3])
```

We can specify the data type of the members of the Ndarray as follows:

```
arr1 = np.array([1, 2, 3], dtype=np.uint8)
```

This URL has a full list of the data types supported by Ndarray:

```
https://numpy.org/devdocs/user/basics.types.html
```

Multidimensional Ndarrays

We can create multidimensional arrays as follows:

```
arr1 = np.array([[1, 2, 3], [4, 5, 6]], np.int16)
arr1
```

The output is as follows:

```
array([[1, 2, 3],
       [4, 5, 6]], dtype=int16)
```

This is a two-dimensional array. We can also create a multidimensional (3D array in the following case) array as follows:

```
arr1 = np.array([[[1, 2, 3], [4, 5, 6]],
                 [[7, 8, 9], [0, 0, 0]],
                 [[-1, -1, -1], [1, 1, 1]]], np.int16)
arr1
```

The output is as follows:

```
array([[[ 1,  2,  3],
        [ 4,  5,  6]],

       [[ 7,  8,  9],
        [ 0,  0,  0]],

       [[-1, -1, -1],
        [ 1,  1,  1]]], dtype=int16)
```

Indexing of Ndarrays

We can address the elements (also called as the members) of the Ndarrays individually. Let's see how to do it with one-dimensional Ndarrays:

```
arr1 = np.array([1, 2, 3], dtype=np.uint8)
```

We can address its elements as follows:

```
print(arr1[0])
print(arr1[1])
print(arr1[2])
```

Just like lists, it follows C style indexing where the first element is at the position of 0 and the nth element is at the position (n-1).

We can also see the last element with negative location number as follows:

```
print(arr1[-1])
```

The last but one element can be seen as follows:

```
print(arr1[-2])
```

If we use an invalid index as follows:

```
print(arr1[3])
```

it throws the following error:

```
-----------------------------------------------------------------------
IndexError                              Traceback (most recent call last)
<ipython-input-24-20c8f9112e0b> in <module>
----> 1 print(arr1[3])

IndexError: index 3 is out of bounds for axis 0 with size 3
```

Let's create a 2D Ndarray as follows:

```
arr1 = np.array([[1, 2, 3], [4, 5, 6]], np.int16)
```

We can also address elements of a 2D Ndarray:

```
print(arr1[0, 0]);
print(arr1[0, 1]);
print(arr1[0, 2]);
```

The output is as follows:

```
1
2
3
```

We can access entire rows as follows:

```
print(arr1[0, :])
print(arr1[1, :])
```

We can also access entire columns as follows:

```
print(arr1[:, 0])
print(arr1[:, 1])
```

We can also extract the elements of a three-dimensional array as follows:

```
arr1 = np.array([[[1, 2, 3], [4, 5, 6]],
                 [[7, 8, 9], [0, 0, 0]],
                 [[-1, -1, -1], [1, 1, 1]]], np.int16)
```

Let's address the elements of the 3D array as follows:

```
print(arr1 [0, 0, 0])
print(arr1 [1, 1, 2])
print(arr1 [:, 1, 1])
```

We can access elements of Ndarrays this way.

Ndarray Properties

We can learn more about the Ndarrays by referring to their properties. Let us learn all the properties with the demonstration. Let us use the same 3D matrix we used earlier:

```
x2 = np.array([[[1, 2, 3], [4, 5, 6]],[[0, -1, -2], [-3, -4, -5]]], np.int16)
```

We can know the number of dimensions with the following statement:

```
print(x2.ndim)
```

The output returns the number of dimensions:

```
3
```

We can know the shape of the Ndarray as follows:

```
print(x2.shape)
```

The shape means the size of the dimensions as follows:

```
(2, 2, 3)
```

We can know the data type of the members as follows:

```
print(x2.dtype)
```

The output is as follows:

```
int16
```

We can know the size (number of elements) and the number of bytes required in the memory for the storage as follows:

```
print(x2.size)
print(x2.nbytes)
```

The output is as follows:

```
12
24
```

We can compute the transpose with the following code:

```
print(x2.T)
```

NumPy Constants

NumPy library has many useful mathematical and scientific constants we can use in programs. The following code snippet prints all such important constants:

```
print(np.inf)
print(np.NAN)
print(np.NINF)
print(np.NZERO)
print(np.PZERO)
print(np.e)
print(np.euler_gamma)
print(np.pi)
```

The output is as follows:

```
inf
nan
-inf
-0.0
0.0
2.718281828459045
0.5772156649015329
3.141592653589793
```

Summary

In this chapter, we familiarized ourselves with Python and IDLE. We learned how to use Python interpreter and how to write Python scripts. We also learned how to execute Python programs. In the next chapter, we will continue our journey with NumPy.

CHAPTER 3

Introduction to Data Visualization

We learned the basics of NumPy in the previous chapter. This chapter is fully focused on learning more functions of NumPy library. We will also learn the basics of Matplotlib library for data visualizations. Just like the previous chapter, we will have a lot of hands-on programming in this chapter too. The chapter covers the following topics:

- NumPy routines for Ndarray creation
- Matplotlib data visualization

After reading this chapter, we will be comfortable with NumPy and Matplotlib.

NumPy Routines for Ndarray Creation

Let's learn a few routines of NumPy for creation of arrays. Let's create a new Jupyter Notebook for this chapter.

The routine `np.empty()` creates an empty array of given size. The elements of the array are random as the array is not initialized.

```
import numpy as np
x = np.empty([3, 3], np.uint8)
print(x)
```

It will output an array with random numbers. And the output maybe different in your case as the numbers are random. We can create multidimensional matrices as follows:

```
x = np.empty([3, 3, 3], np.uint8)
print(x)
```

© Ashwin Pajankar and Aditya Joshi 2022
A. Pajankar and A. Joshi, *Hands-on Machine Learning with Python*, https://doi.org/10.1007/978-1-4842-7921-2_3

We can use the routine np.eye() to create a matrix of all zeros except the diagonal elements of all the zeros. The diagonal has all the ones.

```
y = np.eye(4, dtype=np.uint8)
print(y)
```

The output is as follows:

```
[[1 0 0 0]
 [0 1 0 0]
 [0 0 1 0]
 [0 0 0 1]]
```

We can also set the position of the diagonal as follows:

```
y = np.eye(4, dtype=np.uint8, k=1)
print(y)
```

The output is as follows:

```
[[0 1 0 0]
 [0 0 1 0]
 [0 0 0 1]
 [0 0 0 0]]
```

We can even have the negative value for the position of the diagonal with all ones as follows:

```
y = np.eye(4, dtype=np.uint8, k=-1)
print(y)
```

Run it and see the output.

The function np.identity() returns an identity matrix of the specified size. An identity matrix is a matrix where all the elements at the diagonal are 1 and the rest of the elements are 0. The following are a few examples of that:

```
x = np.identity(3, dtype= np.uint8)
print(x)
x = np.identity(4, dtype= np.uint8)
print(x)
```

The routine `np.ones()` returns the matrix of the given size that has all the elements as ones. Run the following examples to see it in action:

```
x = np.ones((3, 3, 3), dtype=np.int16)
print(x)
x = np.ones((1, 1, 1), dtype=np.int16)
print(x)
```

Let us have a look at the routine arange(). It creates a Ndarray of evenly spaced values with the given interval. An argument for the stop value is compulsory. The start value and interval parameters have default arguments 0 and 1, respectively. Let us see an example:

```
np.arange(10)
```

The output is as follows:

```
array([0, 1, 2, 3, 4, 5, 6, 7, 8, 9])
```

The routine `linspace()` returns a Ndarray of evenly spaced numbers over a specified interval. We must pass it the starting value, the end value, and the number of values as follows:

```
np.linspace(0, 20, 30)
```

The output is as follows:

```
array([ 0.        ,  0.68965517,  1.37931034,  2.06896552,  2.75862069,
        3.44827586,  4.13793103,  4.82758621,  5.51724138,  6.20689655,
        6.89655172,  7.5862069 ,  8.27586207,  8.96551724,  9.65517241,
       10.34482759, 11.03448276, 11.72413793, 12.4137931 , 13.10344828,
       13.79310345, 14.48275862, 15.17241379, 15.86206897, 16.55172414,
       17.24137931, 17.93103448, 18.62068966, 19.31034483, 20.        ])
```

Similarly, we can create Ndarrays with logarithmic spacing as follows:

```
np.logspace(0.1, 2, 10)
```

The output is as follows:

```
array([  1.25892541,   2.04696827,   3.32829814,   5.41169527,
         8.79922544,  14.30722989,  23.26305067,  37.82489906,
        61.50195043, 100.        ])
```

We can also create Ndarrays with the geometric spacing:

```
np.geomspace(0.1, 20, 10)
```

Run this and see the output yourself.

NumPy is a vast topic itself, and it will take many chapters, perhaps a few books to cover it in great detail. The small examples we saw should give you a fair idea about Ndarrays (the previous chapter) and the routines to create them. We will be using NumPy library frequently to store the data throughout this book.

Matplotlib Data Visualization

Matplotlib is a data visualization library. It is an integral part of the Scientific Python Ecosystem. Many other data visualization libraries are just wrappers on Matplotlib. We will be extensively using Matplotlib's Pyplot module throughout this book. It provides MATLAB like interface. Let's begin by writing the demonstration programs. Type in all the following code in the same notebook that we used for the earlier examples.

The following command is known as magic command that enables Jupyter Notebook to show Matplotlib visualizations:

```
%matplotlib inline
```

We can import the Pyplot module of Matplotlib as follows:

```
import matplotlib.pyplot as plt
```

We can draw a simple linear plot as follows:

```
x = np.arange(10)
y = x + 1
plt.plot(x, y)
plt.show()
```

The output is shown in Figure 3-1.

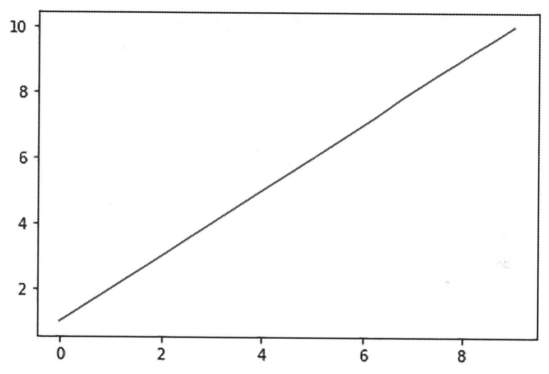

Figure 3-1. *A simple plot*

We can have a multiline plot as follows:

```
x = np.arange(10)
y1 = 1 - x
plt.plot(x, y, x, y1)
plt.show()
```

The output is shown in Figure 3-2.

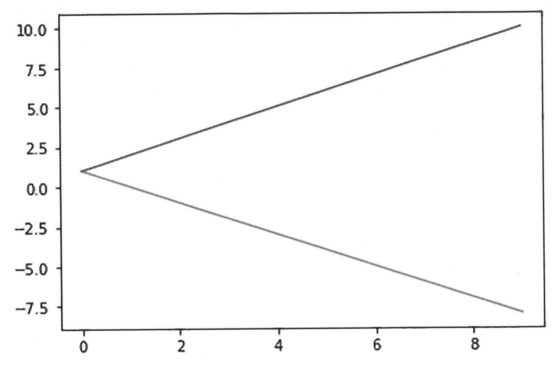

Figure 3-2. *A simple multiline plot*

As we can see, the routine `plt.plot()` can visualize data as simple lines. We can also plot data of other forms with it. The limitation is that it must be single dimensional. Let's draw a sine wave as follows:

```
n = 3
t = np.arange(0, np.pi*2, 0.05)
y = np.sin( n * t )
plt.plot(t, y)
plt.show()
```

The output is shown in Figure 3-3.

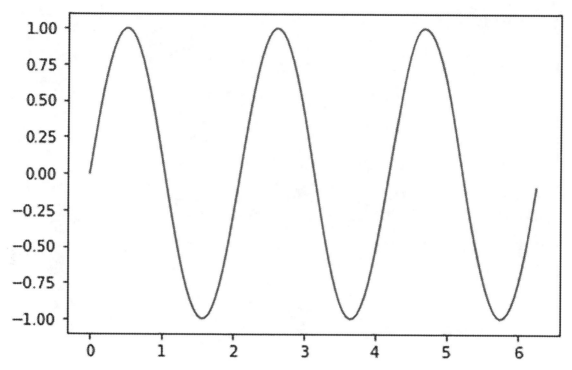

Figure 3-3. *Sinusoidal plot*

We can also have other types of plots. Let's visualize a bar plot.

```
n = 5
x = np.arange(n)
y = np.random.rand(n)
plt.bar(x, y)
plt.show()
```

The output is as shown in Figure 3-4.

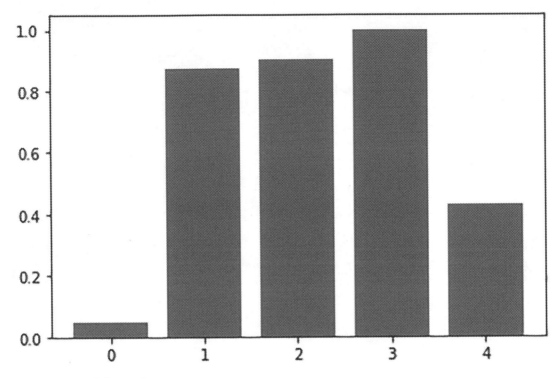

Figure 3-4. *A bar plot*

We can rewrite the same code in an object-oriented way as follows:

```
fig, ax = plt.subplots()
ax.bar(x, y)
ax.set_title('Bar Graph')
ax.set_xlabel('X')
ax.set_ylabel('Y')
plt.show()
```

As we can see, the code creates a figure and an axis that we can use to call visualization routines and to set the properties of the visualizations.

Let's see how to create subplots. Subplots are the plots within the visualization. We can create them as follows:

```
x = np.arange(10)

plt.subplot(2, 2, 1)
plt.plot(x, x)
plt.title('Linear')
```

```
plt.subplot(2, 2, 2)
plt.plot(x, x*x)
plt.title('Quadratic')

plt.subplot(2, 2, 3)
plt.plot(x, np.sqrt(x))
plt.title('Square root')

plt.subplot(2, 2, 4)
plt.plot(x, np.log(x))
plt.title('Log')

plt.tight_layout()
plt.show()
```

As we can see, we are creating a subplot before each plotting routine call. The routine `tight_layout()` creates enough spacing between subplots. The output is as shown in Figure 3-5.

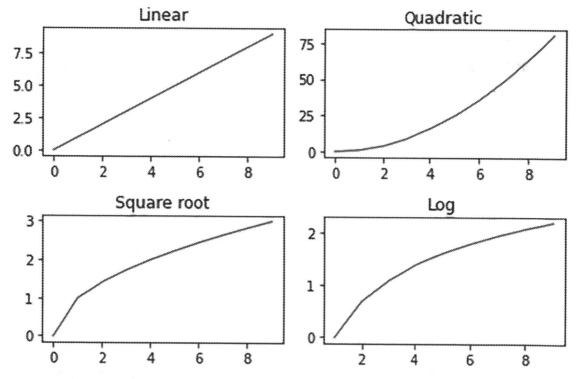

Figure 3-5. *Subplots*

We can write the same code in the object-oriented fashion as follows:

```
fig, ax = plt.subplots(2, 2)

ax[0][0].plot(x, x)
ax[0][0].set_title('Linear')
ax[0][1].plot(x, x*x)
ax[0][1].set_title('Quadratic')
ax[1][0].plot(x, np.sqrt(x))
ax[1][0].set_title('Square Root')
ax[1][1].plot(x, np.log(x))
ax[1][1].set_title('Log')

plt.subplots_adjust(left=0.1,
                    bottom=0.1,
                    right=0.9,
                    top=0.9,
                    wspace=0.4,
                    hspace=0.4)
plt.show()
```

By this time, you must have understood that we can write code either in a MATLAB style syntax or in an object-oriented way.

Let's move ahead with the scatter plot. We can visualize 2D data as scatter plot as follows:

```
n = 100
x = np.random.rand(n)
y = np.random.rand(n)
plt.scatter(x, y)
plt.show()
```

The output is as shown in Figure 3-6.

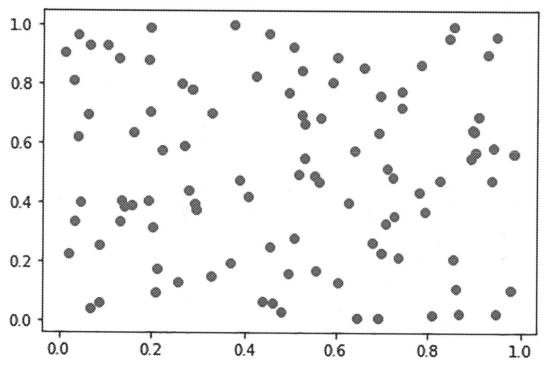

Figure 3-6. *Scatterplot*

The graphical depiction of frequency distribution of any data is known as histogram. We can easily create histograms with Matplotlib as follows:

```
mu, sigma = 0, 0.1
x = np.random.normal(mu, sigma, 1000)
plt.hist(x)
plt.show()
```

Here, mu means mean, and sigma means standard deviation. The output is as shown in Figure 3-7.

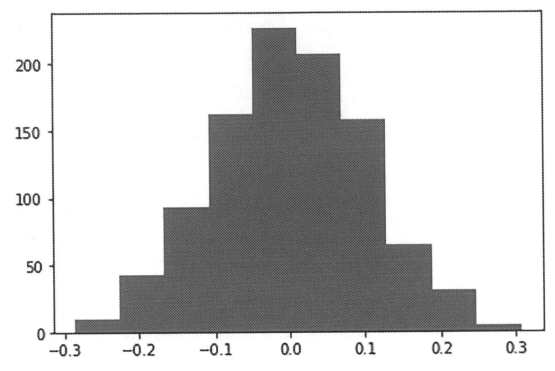

Figure 3-7. *Histogram*

Let's conclude with a pie chart.

```
x = np.array([10, 20, 30, 40])
plt.pie(x)
plt.show()
```

The output is as shown in Figure 3-8.

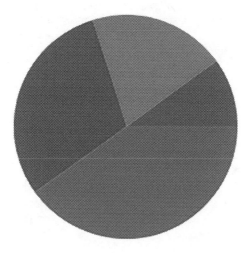

Figure 3-8. *Pie chart*

Summary

In this chapter, we learned a few routines for creation of the NumPy Ndarrays and visualization of the data. NumPy and Matplotlib are important libraries in the Scientific Python Ecosystem, and they are frequently used in Machine Learning programming. These chapters are no way covering all the aspects of NumPy and Matplotlib. However, you must have gained a good understanding of the usage of the routines by the demonstrations in the chapter.

In the next chapter, we will continue our journey with the exploration of the data science library in the Scientific Python Ecosystem, Pandas. We will learn basic data structures, a few operations, and visualization technique.

CHAPTER 4

Introduction to Pandas

We explored the routines for the creation of NumPy Ndarrays and data visualization. We are reasonably comfortable with both libraries, and we can use the routines to create, store, and visualize data with Python programming. In this chapter, we will be acquainted with the data science library of the Scientific Python Ecosystem, Pandas. We will learn the basic data structures, a few operations, and the recipes of visualization with Matplotlib.

The chapter covers the following topics:

- Pandas basics

- Series in Pandas

- Pandas dataframes

- Visualizing the data in dataframes

After reading this chapter, we will be comfortable with the Pandas library of the Scientific Python Ecosystem.

Pandas Basics

Pandas is the data analytics and data science library of the Scientific Python Ecosystem. Just like NumPy, Matplotlib, IPython, and Jupyter Notebook, it is an integral part of the ecosystem.

It is used for storage, manipulation, and visualization of multidimensional data. It is more flexible than Ndarrays and also compatible with it. It means that we can use Ndarrays to create Pandas data structures.

Let's create a new notebook for the demonstrations in this chapter. We can install Pandas with the following command in the Jupyter Notebook session:

```
!pip3 install pandas
```

© Ashwin Pajankar and Aditya Joshi 2022
A. Pajankar and A. Joshi, *Hands-on Machine Learning with Python*, https://doi.org/10.1007/978-1-4842-7921-2_4

The following code imports the library to the current program or Jupyter Notebook session:

```
import pandas as pd
```

I have extensively referred the documentation of Pandas to learn it from scratch by myself. This book covers Pandas basics enough to get started with Machine Learning. This entire chapter is devoted to Pandas. If you wish to learn it in detail, Apress has many books dedicated to the topics. You can also refer to the following URL of Pandas project to learn more:

```
https://pandas.pydata.org/
```

Let us study a few important and essential data structures in Pandas.

Series in Pandas

A Pandas series is a homogeneous one-dimensional array with an index. It can store the data of any supported type. We can use lists or Ndarrays to create series in Pandas. Let's create a new notebook for demonstrations in the chapter. Let's import all the needed libraries:

```
%matplotlib inline
import pandas as pd
import numpy as np
import matplotlib.pyplot as plt
```

Let's create a simple series using list as follows:

```
s1 = pd.Series([1, 2, 3 , 4, 5])
```

If we type the following code:

```
type(s1)
```

we get the following output:

```
pandas.core.series.Series
```

We can also create a series with the following code:

```
s2 = pd.Series(np.arange(5), dtype=np.uint8)
s2
```

The output is as follows:

```
0    0
1    1
2    2
3    3
4    4
dtype: uint8
```

The first column is the index, and the second column is the data column.

We can create a series by using an already defined Ndarray as follows,

```
arr1 = np.arange(5, dtype=np.uint8)
s3 = pd.Series(arr1, dtype=np.int16)
s3
```

In this case, the data type of the series will be considered as the final data type.

Properties of Series

We can check the properties of series as follows.

We can check the values of the members of the series as follows:

```
s3.values
```

The output is as follows:

```
array([0, 1, 2, 3, 4], dtype=int16)
```

We can also check the values of the series with the following code:

```
s3.array
```

The output is as follows:

```
<PandasArray>
[0, 1, 2, 3, 4]
Length: 5, dtype: int16
```

We can check the index of the series:

```
s3.index
```

The following is the output:

```
RangeIndex(start=0, stop=5, step=1)
```

We can check the datatype as follows:

```
s3.dtype
```

We can check the shape as follows:

```
s3.shape
```

We can check the size as follows:

```
s3.size
```

We can check the number of bytes as follows:

```
s3.nbytes
```

And we can check the dimensions as follows:

```
s3.ndim
```

Pandas Dataframes

We can use a two-dimensional indexed and built-in data structure of Pandas known as dataframe. We can create dataframes from series, Ndarrays, lists, and dictionaries. If you have ever worked with relational databases, then you can consider dataframes analogous to tables in the databases.

Let's see how to create a dataframe. Let's create a dictionary of population data for cities as follows:

```
data = {'city': ['Bangalore', 'Bangalore', 'Bangalore',
                 'Mumbai', 'Mumbai', 'Mumbai'],
        'year': [2020, 2021, 2022, 2020, 2021, 2022,],
        'population': [10.0, 10.1, 10.2, 5.2, 5.3, 5.5]}
```

We can create a dataframe using this dictionary:

```
df1 = pd.DataFrame(data)
print(df1)
```

The output is as follows:

```
        city  year  population
0  Bangalore  2020        10.0
1  Bangalore  2021        10.1
2  Bangalore  2022        10.2
3     Mumbai  2020         5.2
4     Mumbai  2021         5.3
5     Mumbai  2022         5.5
```

We can see the first five records of the dataframe directly with the following code:

```
df1.head()
```

Run this and see the output.

We can also create the dataframe with a specific order of columns as follows:

```
df2 = pd.DataFrame(data, columns=['year', 'city', 'population'])
print(df2)
```

The output is as follows:

```
   year       city  population
0  2020  Bangalore        10.0
1  2021  Bangalore        10.1
2  2022  Bangalore        10.2
3  2020     Mumbai         5.2
4  2021     Mumbai         5.3
5  2022     Mumbai         5.5
```

As we can see, the order of columns is different this time.

Visualizing the Data in Dataframes

We have learned the data visualization of NumPy data with the data visualization library Matplotlib. Now, we will learn how to visualize Pandas data structures. Objects of Pandas data structures call Matplotlib visualization functions like plot(). Basically, Pandas provides a wrapper for all these functions. Let us see a simple example as follows:

```
df1 = pd.DataFrame()
df1['A'] = pd.Series(list(range(100)))
df1['B'] = np.random.randn(100, 1)
df1
```

So this code creates a dataframe. Let's plot it now:

```
df1.plot(x='A', y='B')
plt.show()
```

The output is as shown in Figure 4-1.

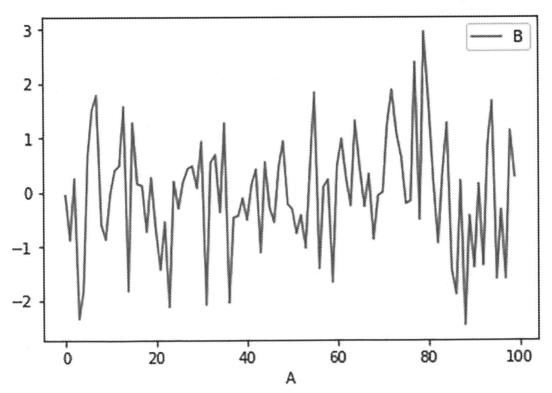

Figure 4-1. *Simple plot*

Now let's explore the other plotting methods. We will create a dataset of four columns. The columns will have random data generated with NumPy. So your output will be definitely different. We will use the generated dataset for the rest of the examples in this chapter. So let's generate the dataset:

```
df2 = pd.DataFrame(np.random.rand(10, 4),
                   columns=['A', 'B', 'C', 'D'])
print(df2)
```

It generates data like below,

```
          A         B         C         D
0  0.191049  0.783689  0.148840  0.409436
1  0.883680  0.957999  0.380425  0.059785
2  0.156075  0.490626  0.099506  0.057651
3  0.195678  0.568190  0.923467  0.321325
4  0.762878  0.111818  0.908522  0.290684
5  0.737371  0.024115  0.092134  0.595440
6  0.004746  0.575702  0.098865  0.351731
7  0.297704  0.657672  0.762490  0.444366
8  0.652769  0.856398  0.667210  0.032418
9  0.976591  0.848599  0.838138  0.724266
```

Let us plot bar graphs as follows:

```
df2.plot.bar()
plt.show()
```

The output is as shown in Figure 4-2.

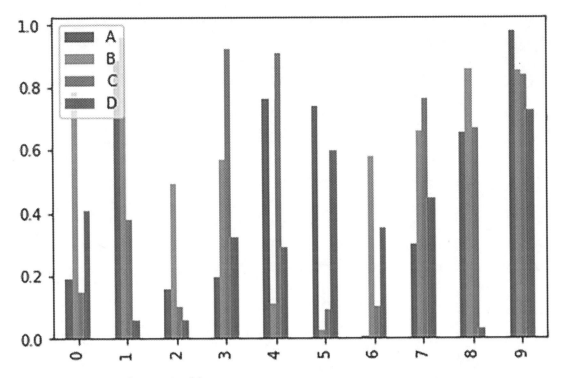

Figure 4-2. *Simple vertical bars*

We can plot these graphs horizontally too as follows:

```
df2.plot.barh()
plt.show()
```

The output is as shown in Figure 4-3.

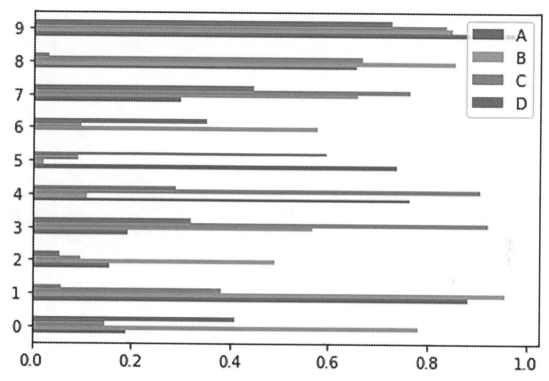

Figure 4-3. *Simple horizontal bars*

These bar graphs were unstacked. We can stack them up as follows:

```
df2.plot.bar(stacked = True)
plt.show()
```

The output is as shown in Figure 4-4.

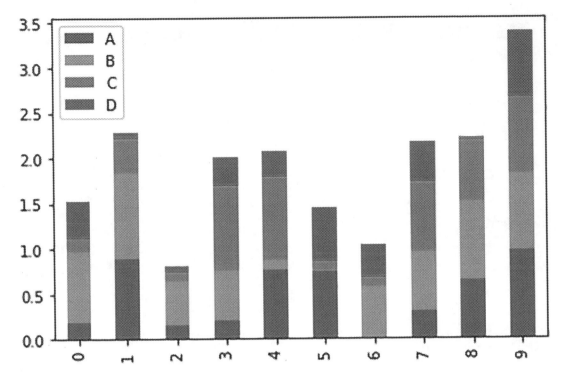

Figure 4-4. *Stacked vertical bars*

We can have horizontal stacked bars as follows:

```
df2.plot.barh(stacked = True)
plt.show()
```

The output is as shown in Figure 4-5.

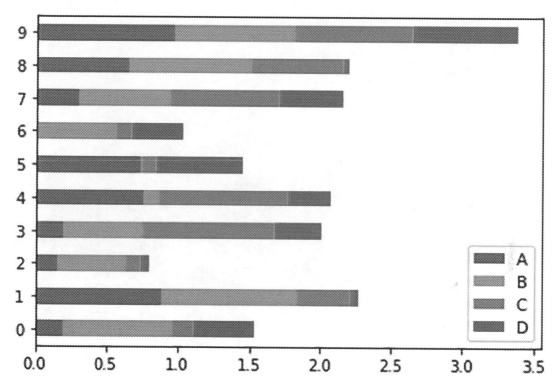

Figure 4-5. *Stacked horizontal bars*

Histograms are a visual representation of the frequency distribution of data. We can plot a simple histogram as follows:

```
df2.plot.hist(alpha=0.7)
plt.show()
```

The output is as shown in Figure 4-6.

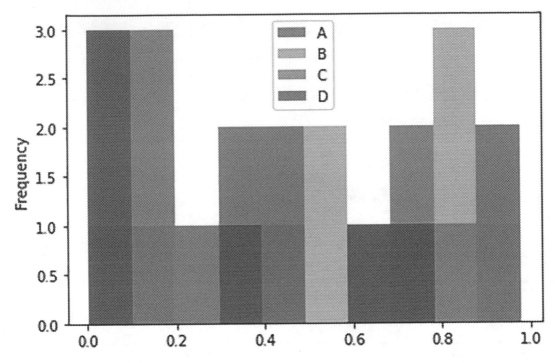

Figure 4-6. *Simple histogram*

We can have a stacked histogram as follows:

```
df2.plot.hist(stacked=True, alpha=0.7)
plt.show()
```

The output is as shown in Figure 4-7.

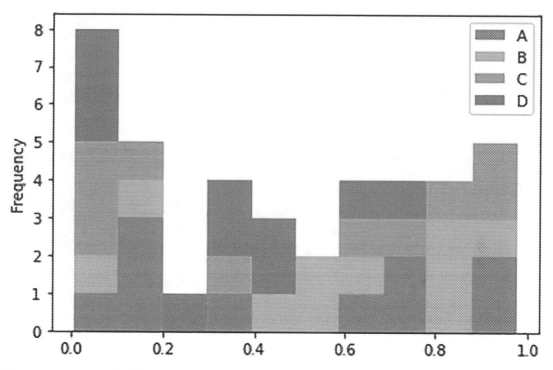

Figure 4-7. *Stacked histogram*

We can also customize buckets (also known as bins) as follows:

```
df2.plot.hist(stacked=True, alpha=0.7, bins=20)
plt.show()
```

The output is as shown in Figure 4-8.

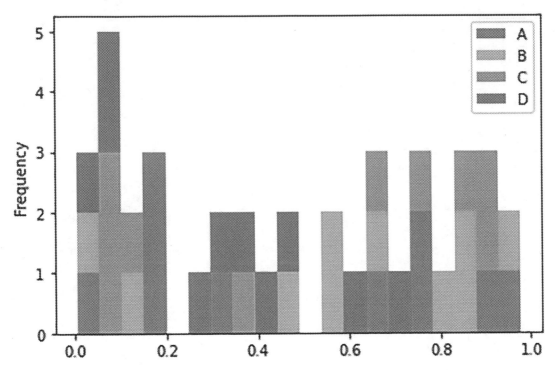

Figure 4-8. *Stacked with custom number of bins/buckets histogram*

We can also draw box plots as follows:

```
df2.plot.box()
plt.show()
```

The output is as shown in Figure 4-9.

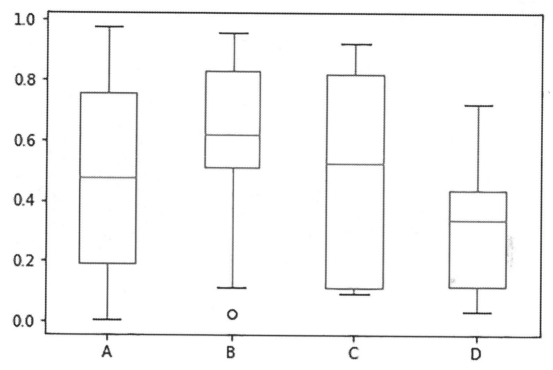

Figure 4-9. *Box plots*

We can draw an area plot as follows:

```
df2.plot.area()
plt.show()
```

The output is as shown in Figure 4-10.

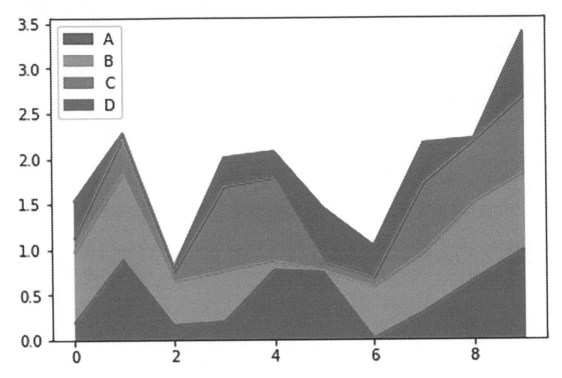

Figure 4-10. *Area plots*

We can draw an unstacked area plot as follows:

```
df2.plot.area(stacked=False)
plt.show()
```

The output is as shown in Figure 4-11.

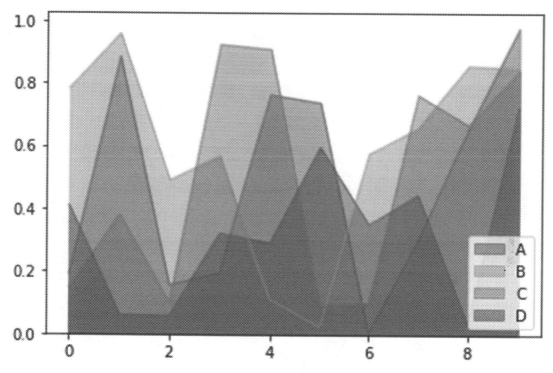

Figure 4-11. *Unstacked area plots*

Summary

In this chapter, we learned the basics of Pandas library and the important data structures, series, and dataframe. We also learned how to visualize the data stored in dataframes. Until now, we have learned the basics of libraries like NumPy, Matplotlib, and Pandas. We will use all these libraries frequently in the demonstrations of the programs for machine learning.

From the next chapter onward, we will dive into the world of machine learning. In the next chapter, we will learn the basics of Scikit-learn.

SECTION 2

Machine Learning Approaches

CHAPTER 5

Introduction to Machine Learning with Scikit-learn

The modern all-connected world is filled with data. According to an estimate, every second, we generate 2.5 quintillion bytes[1] of data around the world.

Machine learning is the science of making computers able to act based on available data without being explicitly programmed to act a certain way. The field of learning from data has been evolving over the past half-century. However, due to the exponential rise in available quantity and quality of data in the past decade, and the available computing resources, we have seen an advent of sophisticated AI agents, self-driving cars, highly intelligent spam filtering, picture tagging applications, personalized web search, and so on. The primary reason for this revolution in technology is being attributed to machine learning.

The most accepted definition of machine learning is given by Tom Mitchell.

> *A computer program is said to learn from experience E with respect to some class of tasks T and performance measure P, if its performance at tasks in T, as measured by P, improves with experience E.*

> —Tom Mitchell

For example, a robotic agent that classifies whether the selected email is a spam or not looks at a well-annotated repository of previous emails that contains the label mentioning whether a given email is a spam or not. This annotated repository acts as the experience for our agent. The task here is a simple binary problem that looks at the input email and produces an output which is yes or no to identify whether an email should be marked as a spam or not. Based on the learning process, the agent will

---

[1] https://techjury.net/blog/how-much-data-is-created-every-day

© Ashwin Pajankar and Aditya Joshi 2022

A. Pajankar and A. Joshi, *Hands-on Machine Learning with Python*, https://doi.org/10.1007/978-1-4842-7921-2_5

classify any new email as a spam correctly or incorrectly. This notion of correctness is captured by the performance measure, P, which here may be simply the ratio of correctly identified spams.

In another example, we can have an agent that plays a game of chess. In the machine learning paradigm, it does not simply look at the chessboard and the positioning of the pieces and simply start recommending the right moves. It observes the thousands of games played in the past that act as the experience while also noticing which moves attributed to a victory. The task is to recommend the next best move in chess. Performance is obtained by observing the percentage of games won against the opponents. Experience is all the moves that were made in the past across multiple games.

In this chapter, we will explore the basics of machine learning and its broad categories and applications. We will follow this with a discussion of Scikit-learn, the most popular library used for machine learning which will be the primary focus of this section of the book. We will begin with installation and understand the top-level architecture and API style that most of the components of Scikit-learn use. We will experiment with a simple example; however, the details will be explained in the relevant future chapters.

Learning from Data

A general notion of machine learning is to focus on the process of learning from data to discover hidden patterns or predict future events. There are two basic approaches within machine learning, namely, supervised learning and unsupervised learning. The main difference is one uses labelled data to help predict outcomes while the other does not.

Supervised Learning

Supervised learning is the set of approaches that requires an explicit label containing the correct expected output for each row in the data through which we want to learn. These labels are either written by hand by domain experts or obtained from previous records or generated by software logs. Such datasets are used to supervise algorithms into classifying the data or predicting outcomes.

Classification

Classification is a suite of supervised learning methods that aim to assign a discrete class label chosen out of a predefined limited set of options (two or more). An example of such labels is a system that monitors the financial transactions and validates each transaction for any kind of fraud. The past data is required, which contains regular transactions as well as fraudulent transactions explicitly stating which data item (or row) is fraudulent. The possible classes are "fraudulent transaction" or "nonfraudulent transaction." Another common example is a sentiment analysis system that takes input text and learns how to classify the given text as "positive," "negative," or "neutral." If you want to build a system to predict movie rating, you will probably collect movie descriptions, tags, genre information and other tags, and the final rating (out of ten), which will act as a label.

Thus, the labels that are assigned are limited and clearly discreet and distinct. In some cases, you can have more than one label to be assigned to one data item – these are called multiclass classification problems.

Regression

Regression is a supervised learning technique that tries to capture the relationship between two variables. We often have one or more independent variables, or the variables that we would always know, and we want to learn how to predict the value of a dependent variable. In regression problems, we want to predict a continuous real value; that means the target value can have infinite possibilities unlike classification that has a selected few. An example of regression is a system to predict the value of a stock the next day based on the value and volume traded on the previous day.

Supervised learning is specifically used in prediction problems where you want to obtain the output for the given data based on past experiences. It is used in use cases like image categorization, sentiment analysis, spam filtering, etc.

Unsupervised Learning

Unsupervised learning is the set of approaches that focus on finding hidden patterns and insights from the given dataset. In such cases, we do not require labelled data. The goal of such approaches is to find the underlying structure of the data, simplify or compress the dataset, or group the data according to inherent similarities.

One common task in unsupervised learning is clustering, which is a method of grouping data points (or objects) into clusters so that the objects that are similar to each other are assigned to one group while making sure that they are significantly different from the items present in other groups. It is highly useful in identifying potential customer segments for directing the marketing efforts and partitioning an image for segmenting into different objects.

Another unsupervised learning approach tries to identify the items that often occur together in the dataset – for example, for a supermarket, it can discover the patterns like bread and milk are often brought together.

Future chapters will cover multiple algorithms for supervised and unsupervised learning.

Structure of a Machine Learning System

A machine learning system either acts by itself or forms a major building block of large-scale enterprise applications.

Out of the production and deployment process and data pipelines, it can be thought of being constructed in two subprocesses in a broad sense. The first part is usually an offline process, which often involves the training, in which we process the real-world data to learn certain parameters that can help predict the results and discover the patterns in previously unseen data. The second part is the online process, which involves the prediction phase, in which we leverage the parameters we have learned before and find the results in previously unseen data.

Based on the quality of results we have obtained as a result, we may decide to make modifications, add more data, and restart the whole process. Such a process, with the help of thorough evaluation metrics, hyperparameter tuning methodologies, and the right choice of features, iteratively produces better and better results.

The whole end-to-end process that is involved can be generalized into a six-stage process outlined here and shown in Figure 5-1:

1. Problem Understanding

2. Data Collection

3. Data Annotation and Data Preparation

4. Data Wrangling

5. Model Development, Training, and Evaluation

6. Model Deployment and Maintenance

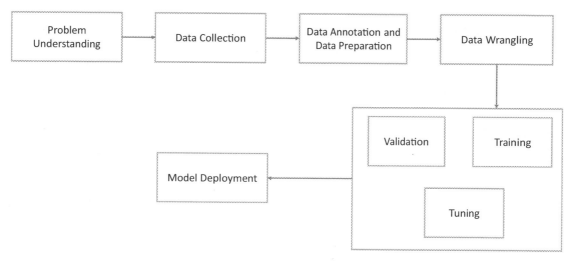

Figure 5-1. *End-to-end process illustrated*

Problem Understanding

Before writing your first line of code, you need to have a thorough understanding of the problem you are trying to solve. This requires discussions with multiple stakeholders and opening the conversations about the right scope of the problems. This would give a clear direction for the next stages of the process.

Data Collection

Once the problem is clear and the right scope has been defined, we can begin to collect the data. Data may come from machine logs, user logs, transaction records, etc. In some cases, the data may be available outright through other teams in the organization or the client itself. In other cases, you might need to hire external agencies or purchase the well-curated data from external providers. In some cases, you might need to collect the data by preparing scraping scripts or leveraging external APIs of some web services.

For many use cases, you can initially search for open source or publicly available datasets that are often shared in public forums or government websites. For the scope of this book, we will restrict ourselves to openly available datasets.

Remember, machine learning is powered by the data. The quality of end results will almost always depend on the quantity and the quality of the data.

Data Annotation and Data Preparation

The raw data that you thus obtain might not always be ready to be used. In case you are working with a supervised problem, you might require a person or a team to assign the correct labels to your data. For example, if you are preparing textual data for sentiment analysis, you may collect it by crawling blogs and social media pages, and after that, you need to take each sentence and assign it a positive or negative label depending on the sentiment polarity it carries.

Data preparation might also require data cleaning, reformatting, and normalization. For example, in the same example, you might want to remove the sentences that are badly structured, or which are too short or in another language. With images, you might require resizing, noise reduction, cropping, etc.

In some cases, you might want to augment the data by combining multiple data sources. For example, if you have official records of an employee's details, you might join another table in your database that contains the employee's past performance records.

Data Wrangling

In all the algorithms that we would study in the future chapters, you will notice that there is an expected format in which they require the data. In general, we want to convert or transform the data from any format into vectors containing numbers. In images, you can look at color channels that contain a value in the 0–255 range for each channel, red, green, and blue. For text, there are multiple common ways to convert the data into vector format. We will study these methods in depth in the next chapter.

Model Development, Training, and Evaluation

In most of the cases, we will leverage existing implementations of algorithms that are already provided in popular packages like Scikit-learn, TensorFlow, PyTorch, etc. However, in some cases, you might need to tweak them before the learning starts. The well-formatted data is then sent to the algorithm for training, during which the model is prepared, which is often a set of parameters or weights related to a predefined set of equations or a graph.

Training usually happens in conjunction with testing over multiple iterations till a model of reliable quality is obtained as shown in Figure 5-2. You will learn the model parameters using a major proportion of the available data and use the same parameters to predict the results for the remaining portion. We will see this process in depth in the future chapters. This is done to evaluate how well your model performs with previously unseen data. Such performance measures can help you improve it by tuning the necessary hyperparameters.

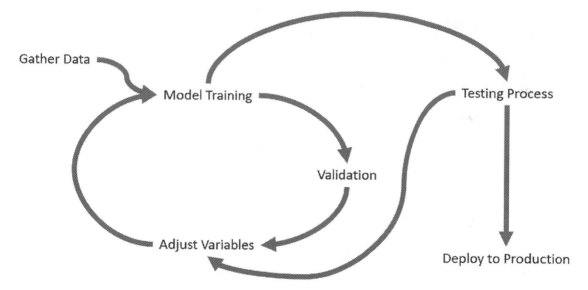

Figure 5-2. *Machine learning training, validation, and testing process*

Once you finalize a set of parameters, you can save your model and use it for inference, which involves utilizing real-world data that follows the pipeline you have prepared in the previous stages.

Model Deployment

Once you have created a model that is ready for inference, you have to make it work as a building block that can be integrated in the production environment. In some cases, this can be a set of scripts that are invoked whenever prediction is required. However, in most enterprise scenarios, this will have to be set up through continuous integration pipelines and hosted in the systems that are capable of handling the anticipated loads.

As the model is deployed, it will see the new data and predict the values for it. In some cases, it might be possible to collect the new data and build an improved version of dataset for future iterations. You will often retrain and update the model from time to time to increase the performance and reliability of the model.

Now that you have a clear idea about the machine learning process, let's discuss one of the most popular packages used for machine learning, Scikit-learn.

Scikit-Learn

Scikit-learn is a highly popular library for machine learning that provides ready-to-use implementations of various supervised and unsupervised machine learning algorithms through a simple and consistent interface. It is built upon the SciPy stack, which involves NumPy, SciPy, Matplotlib, Pandas, etc.

Extensions of SciPy are usually named as SciKits, which provide additional usage for common functionalities above SciPy. Other such common libraries are scikit-image and statsmodels. Scikit-learn is being watched by 46.4k people and forked 21.6k times on GitHub. In this chapter, we will use Scikit-learn, sklearn, and scikit learn interchangeably to refer to this library.

Scikit-learn was initially started in 2007 as a Google Summer of Code project by David Cournapeau, who has also been involved in the development of NumPy and SciPy. By 2010, more developers were starting to get involved, and the first public release was made in February 2010.

Currently, it is one of the most popular libraries for machine learning applications. As of writing of this book, Scikit-learn 0.24 is the latest stable version.

Installing Scikit-Learn

First, check if you have Scikit-learn installed already, probably if you have a distribution installed. Create a new cell in Jupyter Notebook and run the following:

```
import sklearn
```

If you didn't get any message in the output, you already have Scikit-learn installed.

If you do not have Scikit-learn installed, you might get an error that contains the following line: ModuleNotFoundError: No module named 'sklearn1'

In that case, you can install Scikit-learn either from the command line or from within the Scikit-learn code cell. Please note that Scikit-learn requires the basic SciPy stack. Before installing, you must ensure that you have a running Python (3.6 or above for the purposes of this book), compatible with NumPy, SciPy, Joblib, Matplotlib, and Pandas.

Run the following command to install:

```
!pip install scikit-learn
```

You can try running the import statement again, and it should work now. For more details on the installation process, you can refer to the official documentation page:

```
https://scikit-learn.org/stable/install.html.
```

Another alternate highly recommended way is to install a distribution that provides the complete stack required for such projects through an easy-to-configure interface that allows you highly customized virtual environments. One such popular distribution is Anaconda. You may download and install the version suitable for your system from here:

```
www.anaconda.com/products/individual.
```

Apart from the virtual environment manager (conda) and preconfigured installation of required libraries, Anaconda also comes with Jupyter Notebook and other management tools.

Understanding the API

One primary reason for popularity and growth of Scikit-learn is the simplicity of use despite the powerful implementation. It has been designed with simple conventions along the following broad principles (refer to the 2013 paper):

- Consistency, so that the interfaces are similar and limited

- Inspection, to provide transparent mechanisms to store and expose the parameters

- Nonproliferation, to allow only the learning algorithms to be represented using custom classes

- Composition, to allow machine learning tasks to be arranged as simple building blocks

- Sensible defaults, to provide a trustworthy fault baseline value for user-defined parameters

As we will see in the next chapter, machine learning methods expect the data to be present in sets of numerical variables called features. These numerical values can be represented as a vector and implemented as a NumPy array. NumPy provides efficient vectorized operations while keeping the code simple and short.

Scikit-learn is designed in a way to have similar interfaces across the functionalities offered by the library. It is organized around three primary APIs, namely, estimator, predictor, and transformer. Estimators are the core interface implemented by classification, regression, clustering, feature extraction, and dimensionality reduction methods. An estimator is initialized from hyperparameter values and implements the actual learning process in the **fit** method, which you call while providing the input data and labels in the form of X_train and y_train arrays. It will run the learning algorithm to learn the parameters and store them for future use.

Predictors provide a predict method to take the data which needs to be predicted through a NumPy array that we usually refer to as X_test. It applies the required transformation with respect to the parameters that have been learned by the **fit** method and provides the predicted values or labels. Some unsupervised learning estimators provide a predict method to obtain the cluster labels.

Transformer interfaces implement the mechanism to transform the given data in the form of NumPy array through the preprocessing and feature extraction stages. Scaling and normalization methods implement the **transform** method which can be called after learning the parameters. We will discuss transformation in depth in the next chapter.

Several algorithm implementations in Scikit-learn implement one or more of these three interfaces. Some methods can be chained to perform multiple tasks in a single line of code. This can be further simplified with the use of **Pipeline** objects that chain multiple estimators into a single one. Thus, you can encapsulate multiple preprocessing, transformation, and prediction steps into a single object.

```
pipe = Pipeline([('scaler', StandardScaler()), ('svc', SVC())])
```

The object pipe can be used directly to apply scaling and SVM predictor to the data that will be provided later.

Your First Scikit-learn Experiment

Before we dive into machine learning algorithms using Scikit-learn, we can do a short hello-world experiment with Scikit-learn. We will use a simple dataset called **iris** dataset that contains the petal and sepal length of three varieties of iris flowers (Figure 5-3).

Figure 5-3. *The three types of iris flowers in the iris dataset*

Thankfully, Scikit-learn comes with some datasets ready for use. We will import this dataset directly and learn a quick model using it. To load the dataset, enter the following lines:

```
from sklearn import datasets
iris = datasets.load_iris()
print (iris)
print (iris.keys())
```

You should be able to see the components of iris object as a dictionary, which includes the following keys: 'data', 'target', 'frame', 'target_names', 'DESCR', 'feature_names', and 'filename'. Right now, data, target and target_names, and feature_names are of interest to us.

Data contains a 2D NumPy array that contains 150 rows and four columns. **Target** contains 150 items containing the numbers 0, 1, and 2 referring to the target_names, which are 'setosa', 'versicolor', and 'virginica', the three varieties of iris flowers. Feature_names contains the meaning of four columns that are contained in the **data**.

```
print (iris.data[:10])
print (iris.target[:5])
```

This should print first five items from the iris dataset, followed by their labels, or the targets as follows:

```
array([[5.1, 3.5, 1.4, 0.2],
       [4.9, 3. , 1.4, 0.2],
       [4.7, 3.2, 1.3, 0.2],
       [4.6, 3.1, 1.5, 0.2],
       [5. , 3.6, 1.4, 0.2]])
array([0, 0, 0, 0, 0])
```

The four columns represent the feature names that you can obtain by typing

```
iris.feature_names
```

which should print

```
['sepal length (cm)',
 'sepal width (cm)',
 'petal length (cm)',
 'petal width (cm)']
```

These dimensions (in centimeters) are what the four numbers in each row shown previously refer to. The targets are the iris varieties you can get from the following:

```
print (iris.target_names)
```

which is an array of three elements.

```
['setosa' 'versicolor' 'virginica']
```

As the first five elements that we printed have target as 0, they belong to the type setosa, which is the smallest flower in the dataset.

We will now create an estimator with an algorithm called Support Vector Machines (SVM) Classifier that we will study in detail in a dedicated chapter. To initialize and learn the parameters, use the following lines of code:

```
from sklearn import svm
clf = svm.SVC(gamma=0.001, C=100.)
clf.fit(iris.data[:-1], iris.target[:-1])
print (clf.predict(iris.data))
```

In this example, we first imported the Support Vector Machines module from Scikit-learn that contains the implementation of the estimator that we wanted to use in this example. This estimator, svm.SVC, is initialized with two hyperparameters, namely, gamma and C with standard values. In the next line, we instruct the clf estimator object to learn the parameters using the fit() function that usually takes two parameters, input data, and the corresponding target classes. In this process, the SVM will learn the necessary parameters, that is, the boundary lines based on which it can divide the three classes – 0, 1, and 2, referring to Iris setosa, Iris virginica, and Iris versicolor. In the last line, we use the predict() method of the predictor to print the predicted targets for the original dataset according to the model that has been learned. You will see an output like the following:

```
array([0, 0, 0, 0, 0, 0, 0, 0, 0, 0, 0, 0, 0, 0, 0, 0, 0, 0, 0, 0, 0,
0, 0, 0, 0, 0, 0, 0, 0, 0, 0, 0, 0, 0, 0, 0, 0, 0, 0, 0, 0, 0, 0,
0, 0, 0, 1, 1, 1, 1, 1, 1, 1, 1, 1, 1, 1, 1, 1, 1, 1, 1, 1, 1, 1,
1, 1, 1, 1, 1, 2, 1, 1, 1, 1, 1, 2, 1, 1, 1, 1, 1, 1, 1, 1, 1, 1,
1, 1, 1, 2, 2, 2, 2, 2, 2, 1, 2, 2, 2, 2, 2, 2, 2, 2, 2, 2, 2, 2,
2, 2, 2, 2, 2, 2, 2, 2, 2, 2, 2, 2, 2, 2, 2, 2, 2, 2, 2, 2, 2, 2,
2, 2, 2])
```

Note the minor inconsistencies in the predicted results – the couple of records that have been inconsistently marked. These are the predicted results, not the actual targets present in the data.

Summary

In this chapter, we have obtained a top-level picture of AI and machine learning, explored the ML development process, and understood the basics of Scikit-learn API. Other chapters in the next section will discuss machine learning algorithms in detail and discuss the insights on tuning the tools to get the best possible results. We will begin this with the next chapter in which we discuss how the data in various formats needs to be treated before beginning the training process.

## CHAPTER 6

Preparing Data for Machine Learning

As we saw in the previous chapter, data collection, preparation, and normalization is one of the primary steps in any machine learning experiment. Almost all of the machine learning algorithms (there are some interesting exceptions though) work with vectors of numbers. Even the other algorithms or tools would require data to be formatted in a specific way. So regardless of the kind of original data you are working with, you should know how to convert it into a usable format without losing necessary details.

We will begin this chapter by first looking at different kinds of data variables that we expect to find in our experiments. After establishing that distinction, we will get in depth of processing well-structured data from comma-separated values (CSV) files – which would help us understand how to deal with data inconsistencies, missing data, etc., followed by feature selection and feature generation – which will help us convert rows of data into numerical vectors. We will then discuss the standard methods to vectorize text and images, which are the most popular unstructured types of data.

Types of Data Variables

Data variables are often classified based on the arithmetic properties they support. Consider a dataset containing personal details – which will contain a column of the city of the university the person graduated in and the annual salary they are getting in their primary full-time job. Both the columns contain data that might have an impact in the analysis of the predictive systems that will be built.

However, there is a major distinction in terms of what you can do with each column. You might say that person A earns twice that of person B. All the basic operations can be applied on these. But when you look at the city of their university, the only operation you

© Ashwin Pajankar and Aditya Joshi 2022
A. Pajankar and A. Joshi, *Hands-on Machine Learning with Python*, https://doi.org/10.1007/978-1-4842-7921-2_6

can apply is to see whether person A and person B studied in the same city. There are primarily two types of data variables based on this distinction.

1. Continuous variables, which can take any positive or negative real number. These variables are not limited to possible values, and the level of preciseness depends on the method of measurement. For example, your height might be measured as 170 cm. This might be a rounded form of a more precise height, say, 170.110 cm.

2. Discrete variables, which can take only a particular value from the allowed set of possible values. For example, names of cities, level of an employee, etc. Even if these variables take numerical values, they will be limited to a set of possible values rather than a real number in the number line. For example, the block number or sector number in the address of an employee will be treated as a discrete variable.

Another distinction is based on the Scales of Measurement that gives a clearer distinction based on the values a variable may obtain and the arithmetic operations that can be applied on them. In this manner, variables can be thought as being on one of the four scales – nominal, ordinal, interval, or ratio.

Nominal Data

Nominal data is a type of data that can take any arbitrary nonnumerical values. This kind of data can neither be measured nor compared. Nominal data can be qualitative or quantitative; however, due to their nature, no arithmetic operations can be applied on these. For example, names, address, gender, etc., can be regarded as nominal data attributes. The only statistical central tendency that can be studied is the mode, or the quantity that occurs most often in the attribute being studied. Mean and median are meaningless for nominal data.

Ordinal Data

Ordinal data is a type of data whose values follow an order. Though the values don't have a notion of differences and increments, they can be compared as being greater than, lesser than, or equal to each other. Though the values don't support basic arithmetic operations, they can be compared using comparison operators. However, median can be considered as a valid measure of central tendencies. For example, T-shirt sizes (S, M, L, XL, etc.), Likert scale in customer surveys (Always, Sometimes, Rarely, Never), and so on are examples of ordinal data.

Interval Data

Interval data is a type of data that has properties of ordinal data (values can be compared), and the intervals are equally split. These data attributes support addition and subtraction. A good example is the Degrees Fahrenheit scale of temperatures. This means the temperature difference observed between 30°F and 31°F is same as the difference between 99°F and 100°F. However, we can't say that 60°F is twice as hot compared to 30°F.

Ratio Data

Ratio data is a type of data that has a natural zero point and supports all the properties of interval data, along with arithmetic operations of multiplication, division, etc. It has a clear definition of 0 – which denotes none of the attributes being studied. The values are continuous and support all the numeric operations. We can study the statistical measures of central tendencies as well as measures of spread like variation for this kind of data.

Transformation

One of the first steps you would require in any machine learning experiments is to prepare the data and transform it to the form that the algorithm accepts. In some cases, you will need to figure out and extract the required signals and prepare a vector or elements that will represent the data. This process is called feature extraction.

Your attempt for treating any real-world data point will be to convert it into feature vectors. For example, if you have data about students enrolled in a course, each student, the real-world physical entity, can be expressed as their personal demographic information, educational information, etc.

Personal Information: Name, contact number, date of birth, gender, etc.

Educational Information: Highest achieved education level, currently enrolled program, roll number, courses enrolled, marks in each course, etc.

Most of these attributes will need to be transformed. If we choose an algorithm that expects each attribute to be strictly numerical, we might need some of the following transformations to convert nonnumeric features into numeric.

1.   Name: Will be considered meaningless for such approaches. We will need to remove this field.

2.   Date of birth: We can use this field to find the current age of the student.

3.   Gender: Can be one of the various choices. We will see how to process these in the next section.

4.   Highest achieved education level: Can be one of the various choices. Because this is ordinal in nature, we can try a different approach. This will be explained in the following section.

5.   Marks in each course: Can be combined to find the aggregate and bucketed into different grade levels.

Transforming Nominal Attributes

Consider the **Gender** attribute, which may have three values – Male, Female, and Other. This is a nominal attribute as arithmetic operations can't be applied, and we can't compare the values with each other. This can be expressed as a vector of possible values.

Say, there's a student with the following values:
Edward Remirez, Male, 28 years, Bachelors Degree

We can convert the gender column to the set of three values:
Edward Remirez, 0, 1, 0, 28 years, Bachelors Degree

This kind of transformation is called one-hot encoding. Table 6-1 shows how different values of Gender will be transformed in this format. The features that undergo this kind of transformation are converted to binary arrays, where a binary column is created for each possible value of an attribute. If the value in the generated column matches the actual value, it is regarded as 1, or else 0. Scikit-learn provides simple interface for such transformation using `sklearn.preprocessing`.

Table 6-1. *The Field "Gender" Expanded in One-Hot Form*

Gender	Is Female?	Is Male?	Is Other?
Female	1	0	0
Male	0	1	0
Other	0	0	1

Let's prepare a simple dataframe for this example.

```
import pandas as pd

df = pd.DataFrame([["Edward Remirez","Male",28,"Bachelors"],
["Arnav Sharma","Male",23,"Masters"],
["Sophia Smith","Female",19,"High School"]], columns=['Name','Gender','Age',
'Degree'])
```

The dataset contains three rows across four columns as shown in Table 6-2.

Table 6-2. *Dataframe with Three Columns*

	Name	Gender	Age	Degree
0	Edward Remirez	Male	28	Bachelors
1	Arnav Sharma	Male	23	Masters
2	Sophia Smith	Female	19	High School

We need to import the necessary class from `sklearn.preprocessing` and fit an object of `OneHotEncoder`.

```
from sklearn.preprocessing import OneHotEncoder
encoder_for_gender = OneHotEncoder().fit(df[['Gender']])
```

You can verify the values and their column indices using

```
encoder_for_gender.categories_
```

If you have more columns in your dataset, they will be shown as an output to this.

To convert your data into transformed attributes, use the following:

```
gender_values = encoder_for_gender.transform(df[['Gender']])
```

gender_values is a sparse matrix; that is, the elements are stored only where the nonzero values are present. You can convert it to NumPy array using

```
gender_values.toarray()
```

Here, you can see that the gender column of each row has been split into two columns, with the first one representing the Female and second one representing the Male values.

```
array([[0., 1.],
       [0., 1.],
       [1., 0.]])
```

You can repack these values in the dataframe using

```
df[['Gender_F', 'Gender_M']] = gender_values.toarray()
```

The dataframe should now look like Table 6-3.

Table 6-3. *Dataframe with Gender Column Transformed As One-Hot Vector*

	Name	Gender	Age	Degree	Gender_F	Gender_M
0	Edward Remirez	Male	28	Bachelors	0.0	1.0
1	Arnav Sharma	Male	23	Masters	0.0	1.0
2	Sophia Smith	Female	19	High School	1.0	0.0

Transforming Ordinal Attributes

Some attributes have relative ordering of values – these can be transformed in a slightly simpler manner that would also preserve the information about the ordering and help create more meaningful models. Consider the same student:

Edward Remirez, Male, 28 years, Bachelors Degree

We know that the education level follows a pattern of order, that is, High School being at the lower end, followed by Bachelors Degree, Masters Degree, Doctorate Degree, etc. We can assign a numeric value to each label, thus High School being 0, Bachelors as 1, Masters as 2, and Doctorate as 3.

The row can be now written as

Edward Remirez, 0, 1, 0, 28 years, 1

We will continue the experiment with the same dataset as in the previous example. Here's a quick example to convert Degree attribute to Ordinal Encoding.

```
from sklearn.preprocessing import OrdinalEncoder
encoder_for_education = OrdinalEncoder()
encoder_for_education.fit_transform(df[['Degree']])
encoder_for_education.categories_
```

You will see the following categories:

```
[array(['Bachelors', 'High School', 'Masters'], dtype=object)]
```

However, we expect an order as High School > Bachelors > Masters > Doctoral. For this reason, we need to initialize the Ordinal Encoder without expectations of the order of categories. We do the following instead:

```
encoder_for_education = OrdinalEncoder(categories = [['Masters',
'Bachelors','High School', 'Doctoral']])
df[['Degree_encoded']] = encoder_for_education.fit_
transform(df[['Degree']])
```

Now the df transforms as shown in Table 6-4.

Table 6-4. *Dataframe After Encoding Degree Using Ordinal Encoding*

	Name	Gender	Age	Degree	Gender_F	Gender_M	Degree_encoded
0	Edward Remirez	Male	28	Bachelors	0.0	1.0	0.0
1	Arnav Sharma	Male	23	Masters	0.0	1.0	2.0
2	Sophia Smith	Female	19	High School	1.0	0.0	1.0

Table 6-5. *Dataframe Containing Only Numerical Data After Encoding Gender and Degree*

	Age	Gender_F	Gender_M	Degree_encoded
0	28	0.0	1.0	0.0
1	23	0.0	1.0	2.0
2	19	1.0	0.0	1.0

The columns for Gender and Degree can now be removed. We can also remove the column for Name as we believe it to have minimal information that any model should capture.

```
df.drop(columns=['Name', 'Gender', 'Degree'], inplace=True)
```

This dataframe is now ready for use. It should look like the data shown in Table 6-5.

Note   If we have to introduce a new value, say, Post Graduate Diploma, which might stand between Bachelors Degree and Masters Degree, we will need to process the whole data again to reassign the new values.

Normalization

Another important preprocessing step is to normalize the data so that the features are in a similar range. It is highly important especially in case of experiments that use algorithms that are affected by the distribution shape or are based on vector- or distance-based computations. Let's look at the dataframe produced in the previous example. Alternatively, we can create a dataframe that looks exactly the same using the following:

```
df = pd.DataFrame({'Age': {0: 28, 1: 23, 2: 19},
  'Gender_F': {0: 0.0, 1: 0.0, 2: 1.0},
  'Gender_M': {0: 1.0, 1: 1.0, 2: 0.0},
  'Degree_encoded': {0: 0.0, 1: 2.0, 2: 1.0}})
Df
```

This dataframe, as shown in Table 6-6, contains two columns that are entirely ones or zeros, and degree can range from 0 to 4. Age has not been dealt with yet – and for that reason, it can be a number between 16 and 60 in most practical cases. In this section, we will study two primary transformations that can be applied on Age to bring it to a similar range.

Table 6-6. *Recreated Dataframe Containing Only Numerical Data*

	Age	Gender_F	Gender_M	Degree_encoded
0	28	0.0	1.0	0.0
1	23	0.0	1.0	2.0
2	19	1.0	0.0	1.0

Min-Max Scaling

Min-max scaling transforms each feature by compressing it down to a scale where the minimum number in the dataset maps to zero and the maximum number maps to one. The transformation is given by[1]

$$x_{std} = \frac{x - x_{min}}{x_{max} - x_{min}} \quad x_{scaled} = x_{std} * (x_{max} - x_{min}) + x_{min}$$

Feature range (min, max) can be configured if required.

```
from sklearn.preprocessing import MinMaxScaler
scaler = MinMaxScaler()
scaler.fit(df[['Age']])
df['Age'] = scaler.transform(df[['Age']])
```

Now you can verify that the Age column has been resolved to a 0–1 range as shown in Table 6-7.

---

[1]https://scikit-learn.org/stable/modules/generated/sklearn.preprocessing. MinMaxScaler.html

Table 6-7. *Dataframe After Transforming Age Column Using Min-Max Scaling*

	Age	Gender_F	Gender_M	Degree_encoded
0	1.000000	0.0	1.0	0.0
1	0.444444	0.0	1.0	2.0
2	0.000000	1.0	0.0	1.0

Standard Scaling

Standard scaling standardizes the feature values by removing the mean and scaling to unit variance. The value thus represents the z-value with respect to the mean and variance of the column. The standard score of a sample is calculated[2] as

$$z = \frac{(x - \mu)}{s}$$

where μ is the mean and s is the standard deviation of the samples.

We can take the original values of the Age columns and scale it using StandardScaler.

```
from sklearn.preprocessing import StandardScaler
scaler = StandardScaler()
scaler.fit(df[['Age']])
df['Age'] = scaler.transform(df[['Age']])
```

This should lead to the values as shown in Table 6-8.

Table 6-8. *Dataframe After Transforming the Age Column Using StandardScaler*

	Age	Gender_F	Gender_M	Degree_encoded
0	1.267500	0.0	1.0	0.0
1	-0.090536	0.0	1.0	2.0
2	-1.176965	1.0	0.0	1.0

---

[2] https://scikit-learn.org/stable/modules/generated/sklearn.preprocessing.
StandardScaler.html

You can view the parameters of the scaler using

```
scaler.mean_
Out: array([23.33333333])
```

```
scaler.scale_
Out:array([3.68178701])
```

Preprocessing Text

A lot of real-world data is present in the form of text – maybe as comments to a survey, or a product review on an ecommerce website, or as a social media text that we want to leverage. The subject of processing, understanding, and generating text is covered under a major field of study called Natural Language Processing (NLP). In this section, we will explain basic techniques for converting text to vectors that might be required by some of the future examples in this book.

Preparing NLTK

One of the most popular libraries for NLP in Python is NLTK, or Natural Language ToolKit. In case you haven't used NLTK on your machine before, we will need to check if it is ready to use, and if not, we will download the required models.

Run the following:

```
from nltk.tokenize import word_tokenize
```

If the operation is successful, it will not produce any output. If you are using NLTK for the first time, using NLTK functions can lead to an error that might look like the following. This means the pretrained models required for basic operations of NLTK are not present in your system.

```
LookupError:
**********************************************************************
  Resource punkt not found.
  Please use the NLTK Downloader to obtain the resource:

  >>> import nltk
  >>> nltk.download('punkt')

  For more information see: https://www.nltk.org/data.html

  Attempted to load tokenizers/punkt/english.pickle
```

If you see such errors, simply run these lines:

```
import nltk
nltk.download()
```

This will open a window similar to Figure 6-1.

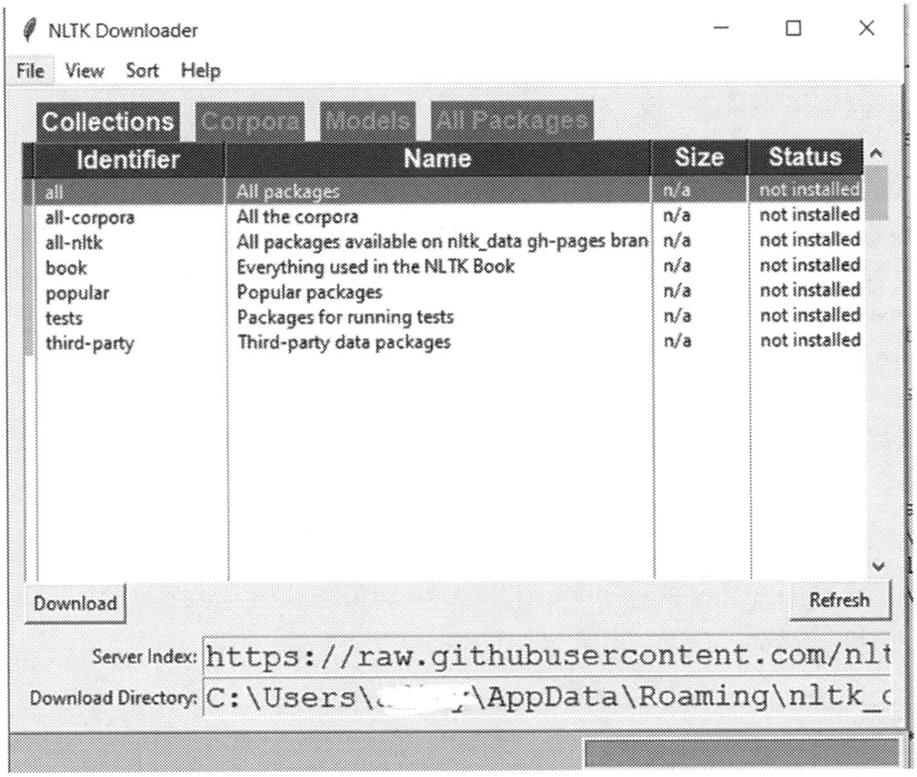

Figure 6-1. *NLTK Package Downloader window*

We can individually select the packages that we need to install. You can only select **book** that will install the required models and tools for examples in the official NLTK book and documentation and include the examples in this book as well. This will take a few minutes, and when finished, the Status column will contain the value as Installed.

Five-Step NLP Pipeline

When you encounter textual data, one of the primary aims should be to develop a pipeline that takes text as input and produces vectors for each sentence or document (or whatever unit of data you wish to consider). It goes through the following steps:

1. Segmentation

Segmentation is the process of finding the sentence boundaries. Each sentence is a fully formed unit that conveys a meaning. In English, a full stop denotes the boundary of a sentence – but it is not always true. For example, full stop or periods can be used in abbreviations – and thus a rule-based approach to consider each full stop as a divider between multiple sentences can be an incorrect assumption. For that reason, there are more complex approaches that can check if the full stop is not a valid sentence separator; or use a complex regular expression pattern to split the sentences. There are more complex and practically usable approaches that leverage Maximum Entropy Models to segment the sentences. In several machine learning experiments, you will be given a sentence directly as a unit of data.

2. Tokenization

Tokenization breaks a sentence or a sequence into individual components or units called tokens. The tokens can be words, special characters, numbers, etc.

Let's try the following in NLTK:

```
from nltk.tokenize import word_tokenize
word_tokenize("Let's learn machine learning")
```

The output is a list containing individual strings representing each token. Each of these can be processed separately.

```
['Let', "'s", 'learn', 'machine', 'learning']
```

In some cases, we wish to perform **Case Folding** so that all the words are present in the same case (preferably, lowercase).

```
tokens = [t.lower() for t in word_tokenize("Let's learn machine learning")]
print (tokens)
['let', "'s", 'learn', 'machine', 'learning']
```

Stemming and Lemmatization

For grammatical reasons, the same word root can be present in different forms in the text. In most cases, they lead to a similar meaning, for example, work, working, worked – all convey a similar meaning in essence, though the interpretation is slightly different. Stemming is the process of extracting the word root.

One popular method used for stemming is called **Porter's Stemmer**.[3] It performs a set of rule-based operations like the following:

```
SSES -> SS
IES  -> I
SS   -> SS
S    ->
```

We can use Porter Stemmer's implementation in NLTK in Python.

```
from nltk.stem.porter import PorterStemmer
stemmer = PorterStemmer()
for token in tokens:
    print(token, " : ", stemmer.stem(token))
```

These lines produce the following output:

```
let  :  let
's  :  's
learn  :  learn
machine  :  machin
learning  :  learn
```

In larger programs, you would prefer the following comprehension:

```
stemmed_tokens = [stemmer.stem(token) for token in tokens]
```

Removing Stopwords

There are several high-frequency words that increase the memory usage but can be ignored with minimal increase in the error. They often add a lot of noise and slow down the processes. These words, like "a," "and," "now," etc., are called stopwords, and NLTK can help us remove them by matching with a word list.

```
from nltk.corpus import stopwords
eng_stopwords = stopwords.words('english')

for token in stemmed_tokens:
    if token in stopwords.words('english'):
        stemmed_tokens.remove(token)
```

---

[3] https://tartarus.org/martin/PorterStemmer/

You can check the complete list of stopwords mentioned in NLTK by printing the values of eng_stopwords. At the time of writing, it contains 179 words. However, these words comprise more than 20% of the English text, thus causing immediate reduction of the size of data.

Preparing Word Vectors

As with all other types of data, text also needs to be converted into a vector form. We have several mechanisms – the simplest being treating a data point (or a sentence) as a bag of words, which can be encoded similar to one-hot mechanism with variation to either put the number 1 in the columns representing all the present words or put the count of the number of occurrences of the word in the given sentence. An example is shown in Table 6-9.

Table 6-9. *Bag of Words Representation of Some Sentences*

	about	bird	heard	is	the	word	you
About the bird, the bird, bird bird bird	1	5	0	0	2	0	0
You heard about the bird	1	1	1	0	1	0	1
The bird is the word	0	1	0	1	2	1	0

For this purpose, we can use CountVectorizer or one of the other Vectorizers available in Scikit-learn. We will see the use of these in the following end-to-end example.

```
from nltk.tokenize import word_tokenize
from nltk.corpus import stopwords
from nltk.stem.porter import PorterStemmer
from sklearn.feature_extraction.text import CountVectorizer

stemmer = PorterStemmer()
eng_stopwords = stopwords.words('english')

data = ["Let's learn Machine Learning Now", "The Machines are Learning",
"It is Learning Time"]
tokens = [word_tokenize(d) for d in data]
tokens = [[word.lower() for word in line] for line in tokens]
```

```
for i, line in enumerate(tokens):
    for word in line:
        if word in stopwords.words('english'):
            line.remove(word)
    tokens[i] = ' '.join(line)

matrix = CountVectorizer()
X = matrix.fit_transform(tokens).toarray()
```

Here, X will become a 2D array of the shape = (m,n) where m is the number of rows in the data and n is the vocabulary size, or the number of unique words that are considered to represent a text vector.

You can visualize the text vectors by comparing feature names with the values:

```
pd.DataFrame(X, columns=matrix.get_feature_names())
```

Table 6-10 shows the final vector representation that can be used as input to machine learning algorithms that we will study in the upcoming chapters.

Table 6-10. *Sentences Expressed as Bag of Words in a Dataframe*

	learn	learning	let	machine	machines	time
0	1	1	1	1	0	0
1	0	1	0	0	1	0
2	0	1	0	0	0	1

Preprocessing Images

Processing images is a large subset of machine learning and computer vision that has been dealt with in separate books. In this section, we would introduce the basic concepts that might be required in some of the examples in this book.

Full-color images can be seen as a three-dimensional array, where two dimensions are used to represent the row and column number of a pixel and the third dimension represents the color channel, red, green, or blue. The value in each cell represents the intensity of each color channel at the given cell.

One simple way to reading an image as a three-dimensional array is to use Matplotlib's imread function. You can put the path to any image located in your machine using the following:

```
import matplotlib.pyplot as plt
img = plt.imread('C:\\Users\\johndoe\\Documents\\ Images\\puppy.jpg')
```

In case of Linux or Mac, please make sure you write the path in the correct format:

```
img = plt.imread('/home/johndoe/Pictures/puppy.jpg')
```

```
(Australian Shepherd Red Puppy by Andrea Stöckel)
```

We can show the image in Figure 6-2 using

```
plt.imshow(img)
```

Figure 6-2. *Imported image displayed using pyplot imshow( )*

The third dimension represents the channel, which might be red, green, or blue. This is how we can get the pixel values for the Red channel in the image, which can be displayed as shown in Figure 6-3:

```
plt.imshow(img[:,:,0])
```

Figure 6-3. *Displaying a single color channel of the image*

You can index any particular pixel of the image using img [row , col , channel].
You can view the results using imshow() as shown in Figure 6-4.

```
cropped_image = img[200:1000,700:1500, :]
```

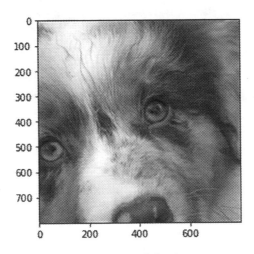

Figure 6-4. *Displaying a cropped section of the image*

```
plt.imshow(cropped_image)
```

There are extensive libraries for image processing and computer vision, the most
popular being OpenCV and Scikit-Image. Here's an example on detecting edges using
Scikit-Image, which produces the output as shown in Figure 6-5.

```
from skimage import io,filters

img = plt.imread('C:\\Users\\johndoe\\Documents\\ Images\\puppy.jpg')
edges = filters.sobel(img)
io.imshow(edges)
io.show()
```

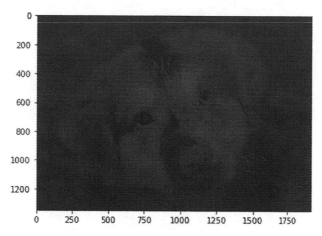

Figure 6-5. *Displaying edges found using Sobel filter*

This filter is used to create an image emphasizing edges of the original image, which can act as a part of the pipeline of a system involving image recognition or classification.

Summary

Using the methods present in this chapter, you should be able to capture your data, transform it to the expected format, and perform necessary scaling before sending it to the next steps down the machine learning pipeline.

In the next chapters, we will study machine learning algorithms and see practical examples on how to use them using Scikit-learn to train models.

CHAPTER 7

Supervised Learning Methods: Part 1

Supervised Learning is the task of learning to predict a numerical or a categorical output for a given input sample. In such problems, you will either obtain or create a dataset with clearly marked outputs. This chapter introduces basic yet important supervised learning methods.

This chapter begins with an explanation of Linear Regression followed by a scikit-learn based experiment. We will also discuss a few measures to determine the quality of regression model. Next, we will discuss a classification method called Logistic Regression, learn a simple model and visualize its predictions in a decision boundary chart. Last, we will learn about Decision Trees which form the basis of a powerful suite of methods that are used for both classification and regression. We will explore a simple example and visualize the tree.

Linear Regression

Linear regression is a supervised learning method specifically to model the relationship between a dependent variable and one or more independent variables. This is an attempt to construct a linear function that outputs the value.

A simple example with one independent variable can be seen in the graph shown in Figure 7-1. Here, we are trying to model the relationship between marks scored by a student and the salary of the first job they got after graduation.

© Ashwin Pajankar and Aditya Joshi 2022
A. Pajankar and A. Joshi, *Hands-on Machine Learning with Python*, https://doi.org/10.1007/978-1-4842-7921-2_7

By a simple look, you can visualize the linear relationship between the two despite several outliers. We attempt to find the line that best justifies all the points of the dataset as a whole.

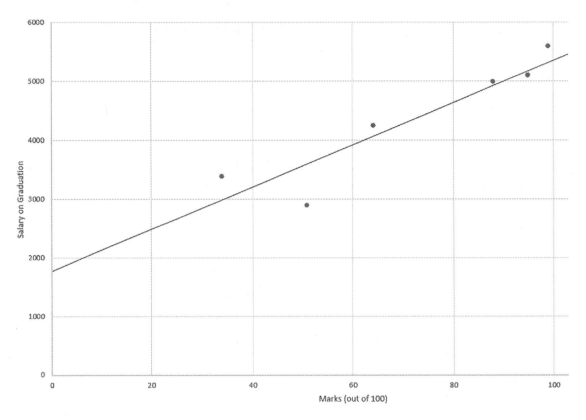

Figure 7-1. *Exploring the relationship between marks of students and salary*

Finding the Regression Line

In linear regression, our aim is to find a function that takes the value (or values) of the independent variable X as input and provides the value of the dependent variable y as output. Basic geometry tells us that an equation of a line is given by

$$y = mx + c$$

where m is the slope of the line and c is where the line meets the y axis. Any line on a 2D plane can be defined by these two parameters.

The aim of learning process or training is to find the best possible values of m and c that try to match the points the most. It is true that due to the nature of real-world data, no line can go through all the points. This leads to the notion of error.

Error is defined as the difference between the independent variable's actual value and the value determined by our regression line. We wish to find the slope m, and y intercept c such that the total cost, given by the average value of squared of the errors.

Usually, the data will have more than one column (components of x) that will be referred as x_1, x_2, x_3... x_n, which will lead to a line that has slopes of m_1, m_2, m_3... m_n across the n axes. Thus, the number of parameters you learn will be (n + 1) where n is the number of columns, or dimensions of the data. For simplicity, we will continue the explanation for a case where you have only one independent column, x.

The slope across each axis is given by

$$b_1 = \frac{\sum_i (x_i - \bar{x})(y - \bar{y})}{\sum_i (x_i - \bar{x})^2}$$

And the y-intercept is

$$b_0 = \bar{y} - b_1 \bar{x}$$

Based on the y-intercept b_0 and one or more sloped b_k, the final equation of the line can be written as

$$y = b_0 + b_1 x_1 + b_2 x_2 + \cdots$$

Linear Regression Using Python

Scikit-learn provides an easy-to-use interface for ordinary least squares implementation.

Let's create a sample dataset to work with. We'll create a Pandas dataframe that contains two columns: an independent variable, which mentions marks of a student out of 100, and a dependent variable, which is the salary they get after graduation. We want to create a linear model that expresses the relationship between the two. We will thus be able to predict the salary a student will get based on the marks they obtained.

```
import numpy as np
import pandas as pd
data = pd.DataFrame({"marks":[34,95,64,88,99,51], "salary":[3400, 2900,
4250, 5000, 5100, 5600]})
```

As we saw in Chapter 5, Scikit-learn has a standard API for most of the common tasks. For learning or fitting the parameters for a model, we can use `fit()` method that takes two arguments: one is the input (independent) variable X, and the other is the output (dependent) variable y. X (usually capital) is a two-dimensional array of shape(n,d) where n is the number of training data rows and d is the number of columns. y (small letter) is a one-dimensional array of shape (n,) that contains one item per each row of training data. To transform data to the correct shape, use the following:

```
X = data[['marks']].values
y = data['salary'].values
```

Mind the extra braces while creating X. This forces Pandas to produce a 2D dataframe of only one column and then convert it to NumPy array. Verify the shape using

print (X.shape, y.shape)

It should produce the following output:

```
(6, 1) (6,)
```

We will now import LinearRegression from Scikit-learn and fit the model.

```
from sklearn.linear_model import LinearRegression
reg = LinearRegression()
reg.fit(X,y)
```

The model has been created. Now we can predict the salary for a student based on their marks using

```
reg.predict([[70]])
```

The output will be a one-dimensional array containing the predicted salary according to the learned model.

```
array([4306.8224479])
```

This also allows you to send an array of multiple rows of data:

```
reg.predict([[100],[50],[80]])
```

which outputs

```
array([5422.45511864, 3563.06733407, 4678.70000481])
```

Visualizing What We Learned

As we know the parameters m and c, we can draw a line to visualize how it fits. We can check the values of coefficient (m) and intercept (c) using the following:

```
print (reg.coef_)
print (reg.intercept_)
Out: [37.18775569]
1703.6795495018537
```

Note that coefficient is an array because we will learn multiple slopes in case your data contains more than one independent variable. Let's import Matplotlib and visualize training data point and regression line.

```
import matplotlib.pyplot as plt

fig,ax = plt.subplots()
plt.scatter(X, y)
ax.axline(  (0, reg.intercept_), slope=reg.coef_ , label='regression line')
ax.legend()
plt.show()
```

The graph it thus generates is shown in Figure 7-2.

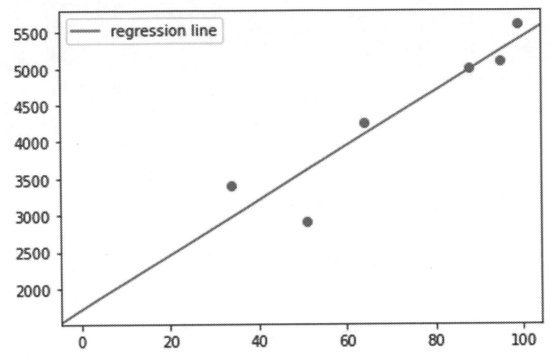

Figure 7-2. *Scatter plot and regression line for marks and salary*

If you are interested, you can make minor changes in this code to print the value of each point in the graph as shown in Figure 7-3.

```
import matplotlib.pyplot as plt
import random

fig,ax = plt.subplots()
fig.set_size_inches(15,7)
plt.scatter(X, y)
ax.axline(  (0, reg.intercept_), slope=reg.coef_ , label='regression line')
ax.legend()
ax.set_xlim(0,110)
ax.set_ylim(1000,10000)

for point in zip(X, y):
    ax.text(point[0][0], point[1]+5, "("+str(point[0]
    [0])+","+str(point[1])+")")

plt.show()
```

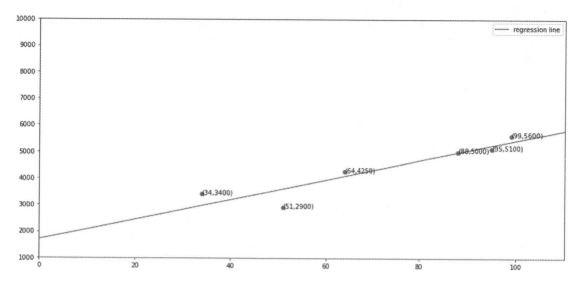

Figure 7-3. *Scatter plot with text labels for each data point*

Evaluating Linear Regression

We have several evaluation measures to check how well our regression model is performing. In this example, we will simply compare how far our predictions are with respect to the actual values in the training dataset. In the following code block, we will prepare a dataframe that compares the actual salary values to the predicted values.

```
results_table = pd.DataFrame(data=X, columns=['Marks'])
results_table['Predicted Salary'] = reg.predict(X)
results_table['Actual Salary'] = y
results_table['Error'] = results_table['Actual Salary']-results_
table['Predicted Salary']
results_table['Error Squared'] = results_table['Error']* results_
table['Error']
```

   results_table should look like that in Table 7-1.

Table 7-1. *Errors for Predicted Salary with respect to Actual Salary*

	Marks	Predicted Salary	Actual Salary	Error	Error Squared
0	34	2968.063243	3400	431.936757	186569.362040
1	51	3600.255090	2900	-700.255090	490357.190739
2	64	4083.695914	4250	166.304086	27657.049103
3	88	4976.202050	5000	23.797950	566.342408
4	95	5236.516340	5100	-136.516340	18636.711137
5	99	5385.267363	5600	214.732637	46110.105415

We can use this table to compute the mean absolute error or mean squared error, or most commonly root mean squared error.

```
import math
import NumPy as np
mean_absolute_error = np.abs(results_table['Error']).mean()
mean_squared_error = results_table['Error Squared'].mean()
root_mean_squared_error = math.sqrt(mean_squared_error)

print (mean_absolute_error)
print (mean_squared_error)
print (root_mean_squared_error)
```

This should produce the three error values:

```
278.9238099821918
128316.12680688586
358.2124045966106
```

Alternatively, you can use internal implementations to achieve the same error values.

```
from sklearn.metrics import mean_squared_error, mean_absolute_
errorprint(mean_squared_error(results_table['Actual Salary'], results_
table['Predicted Salary']))
```

```
print(math.sqrt(mean_squared_error(results_table['Actual Salary'], results_
table['Predicted Salary'])))
print(mean_absolute_error(results_table['Actual Salary'], results_
table['Predicted Salary']))
```

Scikit-learn also provides R Squared value that measures how much variability in a dependent variable can be explained by the model. R Squared is a popularly used measure of goodness of fit of a linear regression measure. Mathematically, it is 1 - sum of squared of prediction error divided by the total sum of squares. It's value is usually between 0 and 1, with 1 representing the ideally best possible model.

```
from sklearn.metrics import r2_score
print ("R Squared: %.2f" % r2_score(y, reg.predict(X)))
```

which gives

```
R Squared: 0.75
```

Logistic Regression

Logistic regression is a classification method that models the probability of a data item belonging to one of the two categories. In the following graph, we wish to predict whether a student will get a job or not based on the marks they obtained in machine learning and in data structures. It is evident that barring a few, the students who had relatively higher score in both machine learning and data structures had got a job offer by the time they graduated. There are two students who had much lesser marks in machine learning, though relatively higher marks in data structures, and they got a job offer.

We want to create a boundary line that separates the students based on their marks on the two subjects so that those who got a job offer by the time they graduated belong to one side of the line and those who didn't on the other side. A potential boundary line is shown in Figure 7-4. Thus, when you get information about a new data point in terms of marks scored in the two subjects, we will be able to predict whether they will get a job or not.

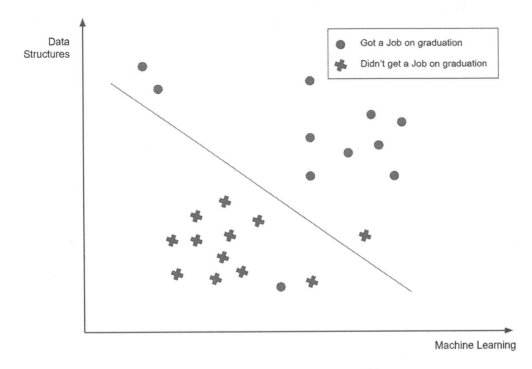

Figure 7-4. *Boundary line in a linear classification problem*

Note   This classification technique is called logistic regression, though it is used to predict binary categorical variables, not continuous like regression methods do. Its name contains the word "regression" because of old conventions as here we try to learn the parameters to regress the probability of a data point belonging to a categorical class.

Line vs. Curve for Expression Probability

Assume that we have a data in only one dimension (say, average marks) and there are two class labels referring to those who got a job and those who didn't. We will call them as positive and negative classes in this discussion. We can try to capture this relationship to find a linear regression line that gives us a probability of a point belonging to a certain class.

However, the target values in the training data are either 0 or 1, where 0 represents the negative class and 1 represents the positive class. However, this kind of data will be hard to capture through a linear relationship. We rather prefer to find a sigmoid or logistic curve that tries to capture the pattern in which most of the predicted values will lie on either y=0 or y=1, and there will be some values within this range. This dependent value can also be treated as the probability for the point to belong to one of the classes.

The sigmoid or logistic function is given as

$$s = \frac{1}{1 + e^{-(\theta_0 + \theta_1 x_1 + \ldots)}}$$

where θ_0, θ_1, … represents the parameter (or parameters, in case of data with more than one column). The S-shaped curve is highly suitable for such use case. The aim of learning process is to find the θ for which we have the minimum possible error of predicting the probability. However, the cost (or error) of the model is based on the predicted class rather than the probability values.

Learning the Parameters

We use a simple iterative method to learn the parameters. Any shift in the values of the parameters causes a shift in the linear decision boundary. We begin with random initial values of the parameters, and by observing the error, we update the parameters to slightly reduce the error. This method is called gradient descent. Here, we try to use the gradient of the cost function to move to the minimum possible cost.

For each parameter, the updated value is given by

Repeat {

$$\theta_j := \theta_j - \alpha \sum_{i=1}^{m} \left(h_\theta \left(x^{(i)} \right) - y^{(i)} \right) x_j^{(i)}$$

} *(simultaneously update for θ_j)*

Here, $h_\theta(x^{(i)})$ is the predicted value for each training row x in the dataset, and $y^{(i)}$ is the actual target class label. $x_j^{(i)}$ is the value of j_{th} column of i_{th} row in the training data. α is an additional parameter (hyperparameter) that is used to control how much the parameter values should be affected in each iteration.

Logistic Regression Using Python

In this example, we will revisit the iris dataset. This dataset contains 150 rows containing sepal length, sepal width, petal length, and petal width of each flower. Based on the sepal and petal dimensions, we want to be able to predict whether a given flower is Iris Setosa, Iris Versicolor, or Iris Virginica.

Let's prepare the dataset using internal datasets provided with Scikit-learn.

```
from sklearn import datasets
iris = datasets.load_iris()
```

load_iris() method returns a dictionary containing the following columns:

```
dict_keys(['data', 'target', 'frame', 'target_names', 'DESCR', 'feature_
names', 'filename'])
```

We can prepare the data for our purpose using the values of these keys. Although we need the raw numbers for training a logistic regression model using Scikit-learn, we will prepare the full dataframe to observe the complete structure of the data in this example.

```
iris_data = pd.DataFrame(iris['data'], columns=iris['feature_names'])
iris_data['target'] = iris['target']
iris_data['target'] = iris_data['target'].apply(  lambda x:iris['target_
names'][x]  )
```

Before we continue this experiment, we will deliberately pick data from Iris Setosa and Iris Versicolor categories to simplify the dataset to be able to fit a binary classification model. There are three types – we will take only two.

```
df = iris_data.query("target=='setosa' | target=='versicolor'")
```

Let's first have a look at the data. There are four variables. Let's pick petal width and length to plot the 100 flowers in 2D.

```
import seaborn as sns
sns.FacetGrid(df, hue='target', size=5).map(plt.scatter, "petal length
(cm)", "petal width (cm)").add_legend()
```

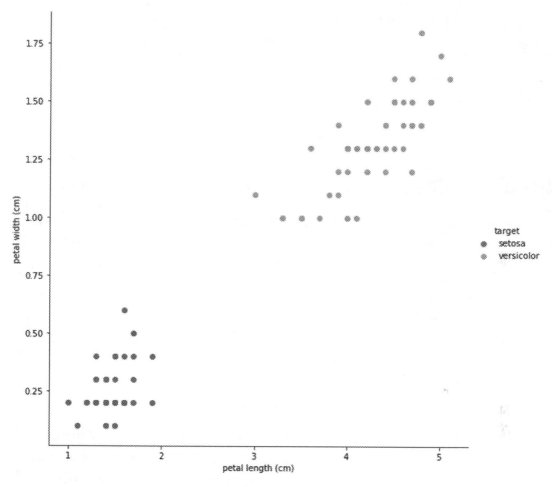

Figure 7-5. *Two types of distinct iris flowers displayed along two dimensions using Seaborn*

As you can see in Figure 7-5, the two categories are mostly distinct, and we should be able to create a boundary that separates the two clearly with almost zero error.

Let's create a logistic regression model using Scikit-learn.

```
from sklearn.linear_model import LogisticRegression
logistic_regression = LogisticRegression()
X = iris_data.drop(columns=['target'])
y = iris_data['target']
logistic_regression.fit(X,y)
```

Let's create a row to test the model:

```
X_test = [[5.6, 2.4, 3.8, 1.2]]
logistic_regression.predict(X_test)
```

And the output of this block is

```
array(['versicolor'], dtype=object)
```

Visualizing the Decision Boundary

To understand how the learned model splits the data into two classes, we will recreate the model using only two dimensions and plot a 2D chart based on sepal length and sepal width. We are limiting the dimensions for easy visualization and understandability.

```
df = iris_data.query("target=='setosa' | target=='versicolor'")[['sepal
length (cm)','sepal width (cm)','target']]
X = df.drop(columns=['target']).values
y = df['target'].values
y = [1 if x == 'setosa' else 0 for x in y]
logistic_regression.fit(X,y)
```

Once we have learned the parameters, we will take every possible point in this 2D space, say, (3.0,3.0), (3,3.1), (3,3.2)…, (3.1,3.0), (3.1,3.1), (3.1,3.2)…, and so on. We will predict the probable class of every such point, and based on the predictions, we will color the point. Eventually, we should be able to see the whole 2D space divided into these two colors, where one color represents Iris Setosa and the other color represents Iris Versicolor.

```
x_min, x_max = X[:, 0].min()-1, X[:,0].max()+1
y_min, y_max = X[:, 1].min()-1, X[:,1].max()+1
xx, yy = np.meshgrid(np.arange(x_min, x_max, 0.02), np.arange(y_min,
y_max, 0.02))

Z = logistic_regression.predict(np.c_[xx.ravel(), yy.ravel()]).
reshape(xx.shape)
plt.rcParams['figure.figsize']=(10,10)
plt.figure()
plt.contourf(xx, yy, Z, alpha=0.4)
```

```
plt.scatter(X[:,0], X[:,1], c=y, cmap='Blues')
plt.xlim(xx.min(), xx.max())
plt.ylim(yy.min(), yy.max())
```

The preceding lines prepare a space of all the points between the minimum possible values of the two columns at the step of 0.2. We then predict the class for each possible point and plot them using contour. We finally overlay the original points colored based on their class label. Here's the boundary you should be able to see.

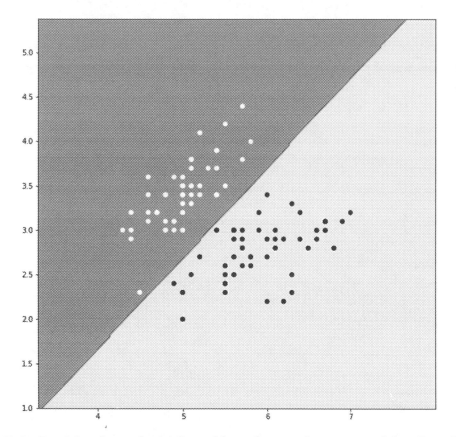

Figure 7-6. *Decision boundary plotted based on performance of the classifier*

The linear boundary shown in Figure 7-6 between the two classes is evident.

This is a simple method but quick and efficient to learn. While using in practice, please note the scale of the features should be comparable, the data should be balanced across the classes, and there shouldn't be too many outliers.

Decision Trees

Decision trees is an effective and highly interpretable suite of machine learning methods that generate a set of rules for regression or classification in the form of a set of decision rules that can be written down in a flowchart-like manner.

A decision tree is drawn upside down with its root at the top. Starting from the root, the tree is full of conditional statements; based on the results of such conditional statements one after the other, you the flow of control will be led down to the leaf nodes, the ones at the end, which denote the target class that is finally predicted.

Consider the following set of conditional tests:

1. Is age > 16?

2. Are marks > 85?

Each data sample goes through a sequence of such tests till it reaches the leaves of the tree, which tree determine the class label based on the proportion of training data samples that fall on them.

Building a Decision Tree

The learning phase of a decision tree algorithm is a recursive process. Within each recursion, it looks at the training data provided for the particular stage and tries to find the best possible split. If there's enough data with enough variation of target classes that can be split leading to a cleaner division of the target labels in the next stage, we proceed with the split. Otherwise, if the training data that's provided to the current stage is too small or belongs to the same target class, we consider it as a leaf node and assign the label of the majority class in the given dataset.

Picking the Splitting Attribute

Picking the one attribute based on which we can split and create a condition thus becomes the core of the algorithm. There are several splitting criterions based on which several implementations are distinguished. One such interesting criterion is to use the concept of entropy, which measures the amount of randomness or uncertainty in the data. The entropy of a dataset depends on how much randomness is present in the node.

In this approach, our aim is to find a splitting criterion that can help us lower the entropy after splitting, thus leading to purer nodes.

Entropy is calculated as $\sum_{i=1}^{c} -p_i \log(p_i)$, where c is the number of possible class labels in the dataset. If a sample is completely homogeneous or belongs to the same class, the entropy is completely zero, and if a sample is equally divided among all the classes, it has an entropy of 1.

To understand this, consider the probability of getting any side in a fair coin. The entropy can be calculated as

$$H(x) = -\sum_{i=1}^{2} \frac{1}{2} \log_2 \frac{1}{2} = 1$$

But if the coin is so heavily biased that we will almost get HEADS, then the entropy will be H(X) = 0.

This means there is no randomness in the data, or we can be 100% certain about the expected outcome of the coin toss. In a decision tree, we want to create the leaf nodes about which we are certain about the class label; thus, we need to minimize the randomness and maximize the reduction of entropy after the split. The attribute we select should lead to the best possible reduction of entropy.

This is captured by the quantity called information gain, which measures how much measure a feature gives us about the class. We thus want to select the attribute that leads to the highest possible information gain.

For data attributes that contain continuous data, we can use Gini index criteria. It is another measurement of impurity. The higher the value of Gini, the higher the homogeneity. CART (Classification and Regression Tree) uses the Gini method to create binary splits.

$$Gini = 1 - \sum_{i=1}^{c} p(i|t)^2$$

While training a decision tree-based classification or regression model using Scikit-learn, we can select the splitting criteria using the criterion hyperparameter.

Decision Tree in Python

In this example, we will use the full Iris dataset that contains information about 150 Iris flower samples across the three categories.

```
import pandas as pd
from sklearn import datasets
iris = datasets.load_iris()
iris_data = pd.DataFrame(iris['data'], columns=iris['feature_names'])
iris_data['target'] = iris['target']
iris_data['target'] = iris_data['target'].apply(  lambda x:iris['target_
names'][x]  )
print(iris_data.shape)
```

Verify the shape of your dataset. It should be

```
(150, 5)
```

Let's separate the features that we will use to learn the decision trees and the associated class labels.

```
X = iris_data[['sepal length (cm)', 'sepal width (cm)', 'petal length
(cm)','petal width (cm)']]
y = iris_data['target']
```

We will then separate data into training and testing dataset.

```
from sklearn.model_selection import train_test_split
X_train, X_test, y_train, y_test = train_test_split (X,y,test_size=0.20,
random_state=0)
```

This will separate the data into 80% training data and 20% testing data. We will thus create the decision trees based on approximately 120 rows (X_train and y_train) thus created. After that, we will predict the results for approximately 30 test data rows (X_test) and compare the predictions with actual class labels (y_test). In the following lines, we initialize a decision tree classifier that uses Gini as the splitting criteria and builds trees up to a maximum depth of 4.

116

```
from sklearn.tree import DecisionTreeClassifier
DT_model = DecisionTreeClassifier(criterion='gini', max_depth=10)
DT_model.fit(X_train, y_train)
```

To find the predictions on the test dataset, we use the following:

```
y_pred = DT_model.predict(X_test)
```

y_pred should be an array containing the predicted target class for each testing data sample. It should look similar to the following though the results might differ due to randomness in how the data is split.

```
array(['virginica', 'versicolor', 'setosa', 'virginica', 'setosa',
'virginica', 'setosa', 'versicolor', 'versicolor', 'versicolor',
'virginica', 'versicolor', 'versicolor', 'versicolor',      'versicolor',
'setosa', 'versicolor', 'versicolor', 'setosa', 'setosa', 'virginica',
'versicolor', 'setosa'], dtype=object)
```

We can evaluate the performance of decision tree by comparing the predicted results with the actual class labels. Accuracy compares what ratio of y_pred is exactly the same as y_test. This should output a number from 0 to 1, with 1 representing 100% accurate results.

```
from sklearn.metrics import accuracy_score
print (accuracy_score(y_test, y_pred))
```

Confusion matrix shows the cross-tabulated count of actual labels and predicted labels.

```
from sklearn.metrics import confusion_matrix
confusion_matrix(y_test, y_pred)
```

This will print a 3x3 grid. First row contains the counts of the test dataset items of first class predicted as each potential target class. First row in the following output example shows that seven Iris Setosa samples are labelled as Iris Setosa, 0 as Iris Versicolor, and 0 as Iris Virginica.

```
Out[14]:
array([[ 7,  0,  0],
       [ 0, 11,  0],
       [ 0,  0,  5]], dtype=int64)
```

We will discuss more evaluation and tuning methods thoroughly in the next chapter.

Pruning the Trees

Decision trees can be made more efficient using pruning, i.e., removing the branches that make the features having low importance. We can assign predefined hyperparameters to perform pruning using the operations like limit the depth of the decision trees, minimum samples that must be present to split an internal node, and minimum samples each leaf should contain.

Scikit-learn provides various implementations for decision trees under tree package. We'll use DecisionTreeClassifier. This package also includes implementation for DecisionTreeRegressor and two more classes that implement extremely randomized tree classifier and regressor.

```
from sklearn.tree import DecisionTreeClassifier
```

Interpreting Decision Trees

One of the qualities of decision trees is that they are easy to understand, interpret, and visualize. If you can view how the decision conditions are arranged and how is the final label assigned, you can easily understand the underlying pattern of the data and the model that has been generated.

Scikit-learn provides an option to export a decision tree in DOT format, which is a popular format to store and share information about the structure and visual properties of graphs or trees. We can use that in conjunction with another library called PyDotPlus that will allow us to create the graph from DOT data. We can alternatively use DOT utility directly to visualize or export the graphs.

To view the trees directly in the Jupyter Notebook, we will also need Graphviz, which is an open source graph visualization software. You may download and install[1] the right distribution for your system. You will also need to install PyDotPlus and Python-Graphviz to extract the information about the decision trees and convert them into a format that can be displayed on Jupyter Notebook.

You can install the requirements using the following:

```
%pip install pydotplus
%pip install python-graphviz
```

---

[1] https://graphviz.org/download/

Once you have a tree created, use the following code block to import required elements:

```
Import IPython
from sklearn.tree import export_graphviz
import pydotplus
import matplotlib.pyplot as plt
dot_data = export_graphviz(DT_model, feature_names= selected_cols)
graph = pydotplus.graph_from_dot_data(dot_data)
img = IPython.display.Image(graph.create_png())
```

In the preceding code, export_graphviz first extracts the structure of the decision tree in DOT format. DOT format[2] uses a simple and standardized syntax to represent graphs and trees using well-defined keywords and symbols. PyDotPlus then reads the DOT format and converts into an object of pydotplus.graphviz.dot that can be then used to create an image and displayed using IPython.display that would look similar to Figure 7-7.

```
IPython.display.display(image)
```

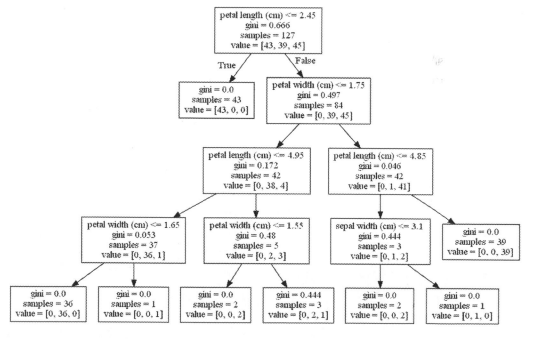

Figure 7-7. *Displaying the decision tree generated after training on Iris dataset*

---

[2] https://graphviz.org/doc/info/lang.html

Each internal node of the tree shows a condition on one of the parameters, based on which we traverse either the right or the left subtree. The leaf node contains a value object that contains three elements. This shows how many training samples ending up on this leaf belong to each class. For example, the leftmost leaf contains value = [0,36,0], which means 36 training samples belong to Iris Versicolor, and none belong to either Iris Setosa or Iris Virginica. Thus, if any sample ends up on this leaf during prediction, it will be predicted as the majority class, which here is Versicolor.

Decision trees can be used for classification as well as regression. If you want to construct a regression tree, you can use `sklearn.tree.DecisionTreeRegressor` and prepare `y_train` as an array (or series) of continuous values instead of class labels.

Summary

In this chapter, we have learned regression and classification methods, namely, linear regression, logistic regression, and decision trees. These are some of the basic supervised learning methods you can use to create a prediction system. We will explore these in depth and discover various ways to more thoroughly evaluate these models and tune these to attain the best possible methods.

CHAPTER 8

Tuning Supervised Learners

As we saw in the previous chapter, there are multiple suites of supervised learning algorithms that can be used to model prediction systems through a labelled training data which might predict a real number (in regression) or one or more discrete classes (in classification). Each method provides a set of features that can be modified or tuned to manipulate the capabilities of the model – which might have a significant effect on the qualities of results thus achieved.

In this chapter, we will study how a machine learning experiment can be designed, evaluated, and tuned to achieve the best possible model.

Training and Testing Processes

A machine learning experiment is divided into two primary phases. The model is first fit on a training dataset. The training dataset contains training tuples that contain an input vector and the corresponding output. The predicted quantity is usually called the target. In this phase, we would first work on improving the signals from the (in case of supervised learning) labelled datasets through feature engineering and then learn the parameters of the model.

In the second part, the model is used to predict the targets for another labelled dataset called test dataset. It also contains the training tuples consisting of an input vector and the output. However, this data is not exposed to the learning algorithm during training – thus, it is unseen by the model. This provides a way to perform unbiased evaluation of the model. We can further modify the process or tweak the algorithm to make predictions that yield better evaluation metrics. The block diagram in Figure 8-1 explains how the training and testing processes occur in a machine learning project.

© Ashwin Pajankar and Aditya Joshi 2022
A. Pajankar and A. Joshi, *Hands-on Machine Learning with Python*, https://doi.org/10.1007/978-1-4842-7921-2_8

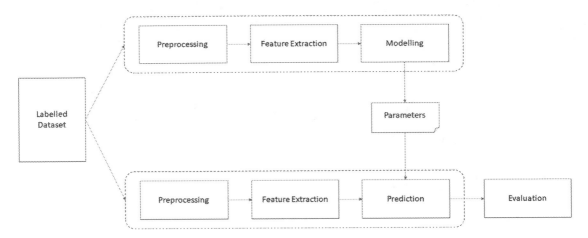

Figure 8-1. *Training and testing processes*

In some cases, we instead use a validation set – which is used to tune the model. Once we've got results that are reliable enough, we then use another test dataset (also called holdout dataset), which returns the final evaluation metrics of the model. In some other experiments, we rather combine all the labelled dataset and use cross validation to come up with the metrics instead. We will discuss this in detail in a later section.

Measures of Performance

Once we create a machine learning model and fit it in a data pipeline to predict results for a previously unseen sample of data, we need to ensure that the model is accurate.

Measuring the quality of results that a classifier model can generate is an important topic that requires sufficient understanding of the model as well as the domain your problem is based on.

Confusion Matrix

Confusion matrix is a simple contingency table that is used to visualize the performance of a classification algorithm which may classify the elements into two or more classes. In the table, each row represents the items belonging to the actual classes, and each column represents the items belonging to the predicted classes.

Let's assume there's a model that captures the diagnostics information of a person and, based on that, predicts whether the person has a viral disease or not. Say, we have

100 labelled rows that we use for testing. Remember, these rows haven't been used to train the model and will be used only to evaluate. The two rows would represent the people who are negative for the virus and those who are positive. The two columns represent the people whom our classification model predicts as being negative or positive. Say, the table for a particular model looks like Table 8-1.

Table 8-1. *Confusion Matrix Example*

		Predicted	
		Negative	Positive
Actual	Negative	90 (True Negative)	3 (False Positive)
	Positive	2 (False Negative)	5 (True Positive)

A quick glance at this table shows that there are a total of 100 items in the test dataset.

Out of them, there are seven (2+5) people who are actually positive for the viral disease, and 93 (90+3) people who are negative.

However, the model has predicted 92 people as negative, including two who are actually positive. The samples that are correctly labelled as negative are called True Negatives (TN), and the ones that are incorrectly labelled as negative are called False Negatives (FN). Similarly, the model predicted eight people as positive, out of which five are correctly predicted as positive, thus indicating the True Positives (TP). The three items that are incorrectly labelled as positive but are actually negative are called False Positives (FP).

True Positives and True Negatives amount to the overall accuracy of your model. False Positives are often called Type 1 Error, and False Negatives are called Type 2 Error. Though there's usually a trade-off between the two, which error should you be more concerned about depends on the problem you're trying to solve.

Recall

Recall is a measure that indicates the ratio of positive test data items that are correctly identified out of all the items that are actually positive. It can be computed from the contingency matrix as

$$Recall = \frac{True\ Positives}{True\ Positives + False\ Negatives}$$

In the preceding example, recall is 5/7 = 0.714

Precision

Precision is the measure that indicates ratio of the number of correctly predicted positive points to the number of all the points that were predicted as positive. It can be computed as

$$Precision = \frac{True\ Positives}{True\ Positives + False\ Positives}$$

In the preceding example, precision will be 5/8, or 0.625. Note that precision and recall don't have units.

Both precision and recall should be as high as possible. It is possible to tune your model to have a better recall and precision by manipulating the hyperparameters of the model as we will see in a later section. However, in some cases, we find as we attempt to increase the recall the precision might relatively decrease. This means we want to capture as many positive samples as possible, though doing this we will also need to tolerate the negative samples that the model will assign as positive. If we try to ensure high precision of the positively predicted samples, we might do that at the cost of reducing the recall.

Accuracy

Accuracy is a simple measure that denotes how many items are correctly classified into both the classes. Here, we have identified 90 negative samples and five positive samples correctly. Thus, the accuracy is (90+5)/100, or 0.95, or 95%.

F-Measure

F-measure or F1-Score is a score obtained by taking the harmonic mean of precision and recall to give a general picture of the goodness of the classification model. Harmonic mean, instead of using the arithmetic mean, penalizes the extreme values more and moves more toward the lower value of the two. It is computed as

$$F - Measure = \frac{2 * Precision * Recall}{Precision + Recall}$$

In this example, F-measure is 2x0.714x0.625/(0.714+0.625), or 0.667.

Performance Metrics in Python

Scikit-learn provides three APIs for evaluating the model quality, namely, estimator score method, scoring parameter, and metric functions. Estimator score method is the model.score() method that can be called for every object of any classifier, regression, or clustering classes. It is implemented within the estimator's code and doesn't require importing any additional modules. It's implementation is different for each estimator. For example, it returns the coefficient of determination R^2 of the prediction. For logistic regression or decision trees, score() returns the mean accuracy on the given test data and labels.

We will now explore the metric functions that are provided under sklearn.metrics.

For the code in this section, we assume that you have created a classifier for Iris flower classification in the previous chapter. Once you have imported the dataset and prepared X_train, y_train, X_test, and y_test objects in the correct format, you can initialize and fit any classifier. We will continue with the decision tree model from the previous chapter that divides the data into one of the three Iris classes.

```
DT_model = DecisionTreeClassifier(criterion="entropy", max_depth=3)
DT_model.fit(X_train, y_train)
```

Now we will import sklearn.metrics so that we can access all the metric functions in this module.

```
import sklearn.metrics
```

Assuming you have trained the model, we will now find predicted class labels for the test dataset.

```
y_pred = DT_model.predict(X_test)
print (y_pred)
```

This should print the array containing predicted values of all the test data samples.

```
['versicolor' 'setosa' 'virginica' 'versicolor' 'virginica' 'setosa'
 'versicolor' 'virginica' 'versicolor' 'versicolor' 'virginica' 'setosa'
 'setosa' 'setosa' 'setosa' 'versicolor' 'virginica' 'versicolor'
 'versicolor' 'virginica' 'setosa' 'virginica' 'setosa' 'virginica'
 'virginica' 'virginica' 'virginica' 'virginica' 'setosa' 'setosa' 'setosa'
 'setosa' 'versicolor' 'setosa' 'setosa' 'virginica' 'versicolor' 'setosa']
```

Print the confusion metrics using

```
sklearn.metrics.confusion_matrix(y_test, y_pred)
```

which should print the confusion metrics. As our data contains three class labels for the three types of Iris flowers, there are three rows representing actual class labels and three columns representing the predicted class sample.

```
array([[15,  0,  0],
       [ 0, 10,  1],
       [ 0,  0, 12]], dtype=int64)
```

From this chart, it can be interpreted that there are 15 samples that belong to the first type (Setosa), which have been correctly classified. However, there are 11 samples in the second type, out of which ten have been correctly classified, and one has been classified as the third type. Then there are 12 samples, all of which are correctly identified as the third type.

sklearn.metrics also contains functions for precision, recall, and F-measure. All these functions take at least two arguments: the actual class labels and the predicted class labels. If you have more than two classes, you can give the additional function parameter for average, which may contain one of the following values:

> binary: Default, the function only reports the results for the positive class labels.

> micro: Calculate metrics globally by counting the total true positives, false negatives, and false positives.

> macro: Calculate metrics for each label, and find their unweighted mean. This does not take label imbalance into account.

> weighted: Calculate metrics for each label, and find their average weighted by the number of instances for each label in the test data and counts for the label imbalance.

We can now find the performance metrics discussed in this section at macro level.

```
p = sklearn.metrics.precision_score(y_test, y_pred, average='micro')
r = sklearn.metrics.recall_score(y_test, y_pred, average='micro')
f = sklearn.metrics.f1_score(y_test, y_pred, average='micro')
a = sklearn.metrics.accuracy_score(y_test,y_pred)

print ("Here're the metrics for the trained model:")
print ("Precision:\t{}\nRecall:\t{}\nF-Score:\t{}\nAccuracy:\t{}".
format(p,r,f,a))
```

This should produce the following output that summarizes the metrics for the model based on which y_pred is provided.

```
Here're the metrics for the trained model:
Precision:    0.9736842105263158
Recall:       0.9736842105263158
F-Score:      0.9736842105263158
Accuracy:     0.9736842105263158
```

Classification Report

Classification report gives most of the important and common metrics required for classification tasks in one single view. It shows the precision, recall, and f-score for each class along with the support, or the number of actual testing samples that belong to the class.

sklearn.metrics.classification_report() returns a formatted string that contains the summary of the scores for each class in the test data. Print it using the following:

```
print (sklearn.metrics.classification_report(y_test, y_pred))
```

This should print the following summary:

	precision	recall	f1-score	support
setosa	1.00	1.00	1.00	15
versicolor	1.00	0.91	0.95	11
virginica	0.92	1.00	0.96	12

accuracy			0.97	38
macro avg	0.97	0.97	0.97	38
weighted avg	0.98	0.97	0.97	38

This presents a clearer picture of the quality of the results that are obtained by predicting based on our model.

Cross Validation

We saw in the previous sections that while approaching a supervised learning problem, we divide the labelled dataset into two components – namely, training set and validation (or testing) set. Rather than relying on a static part of the data for learning the model and using other static part for validation, it is a good idea to come with a rotation of training and testing parts to be able to determine how well will the model generalize to an independent dataset.

In this scheme, we have a predefined ratio of test dataset, say, 25%. That means we divide the dataset into four parts, thus leading to fourfold cross validation (Figure 8-2).

In the first iteration of cross validation experiment, we consider the first part (or first fold) of the labelled data as the test data and the remaining three parts as the training data. We learn the model and then use it to predict the labels for the test data for this iteration, using which, we can compute the accuracy or other measures that we saw in the previous pages.

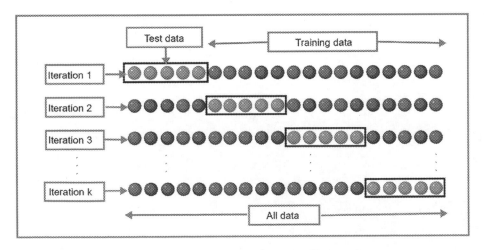

Figure 8-2. *Assigning data points for k-fold cross validation*

After the first iteration is over, we assign the second part (fold) as the test data and the remaining three as training data to create the model. We then attain another set of metrics. We repeat the process k number of times, where k is the number of folds that were thus created. This process is called k-fold cross validation.

Over the k iterations, we will obtain k metrics, which can be averaged to find a more generalizable metric that can be used to tune the hyperparameters.

If k-fold cross validation is set with k equal to the number of data samples, the model is trained on all data points except 1, and prediction is made only for that point. It is called **leave-one-out cross validation**. It is computationally expensive due to its nature.

Another variation is **stratified cross validation**. Instead of creating the k folds randomly, the data is divided by stratified sampling. Stratified sampling divides the data while into training and testing sets by taking the balance of representation of class labels into consideration and is more practical in case of data having nonbalanced class labels.

Why Cross Validation?

If we tweak a model based on a static test set, it is possible to overoptimize and overfit on the test set which may not generalize to more unseen data. The knowledge about the test set can indirectly creep into the model, and evaluation metrics are not generalized. Thus, multiple folds of the data provide an opportunity to not tune the results on one static set.

Cross Validation in Python

For cross validation, we will not consider randomly created train-test splits as in previous examples but work with the features and the values. Let's restart the exercise by loading the data:

```
import pandas as pd
from sklearn import datasets
iris = datasets.load_iris()
X = pd.DataFrame(iris['data'], columns=iris['feature_names'])
y = iris['target']
```

The complete dataset is now present in features X and labels y.

We will now use fivefold cross validation to create multiple splits:

```
from sklearn.model_selection import KFold
kf = KFold(n_splits=5)
kf.get_n_splits(X)
```

This initiates fivefold cross validation that will produce five splitting iterations, each containing approximately 120 elements in training set and 30 elements in testing set. We can look at the index of the elements that are chosen for training and testing in each iteration as follows:

```
for i, (train_index, test_index) in enumerate(kf.split(X)):
    print ("Iteration "+str(i+1))
    print("Train Indices:", train_index, "\nTest Indices:", test_
    index,"\n")
```

This will print the indices of the dataset that are chosen for training dataset and testing dataset in each iteration. The following output is truncated for brevity.

```
Iteration 1
Train Indices: [ 30  31  32  33  34  35  36  37  38  39  40  41  42  43
44  45  46  47 48  49  50  51  52  53  54  55  56 ....
... 136 137  138 139 140 141 142 143 144 145 146 147 148 149]
Test Indices: [ 0  1  2  3  4  5  6  7  8  9 10 11 12 13 14 15 16 17 18 19
20 21 22 23  24 25 26 27 28 29]
Iteration 2
Train Indices: [ 0  1  2  3  4  5  6  7  8  9 10 11 ...
```

Thus, in each split iteration, five in total, k-fold split will indicate the indexes of the elements that should be present in training or testing dataset.

We can use the data points in each iteration to fit the model:

```
score_history = []
for train, test in kf.split(X, y):
    clf = DecisionTreeClassifier()
    clf.fit(X.values[train,:], y[train])
    score_history.append(clf.score(X.values[test,:], y_pred))
```

Alternatively, we can also use cross_val_score from sklearn.model_selection to let the Scikit-learn automatically create multiple models iteratively and present the accuracy metrics to you. We will see these in the examples in the upcoming sections.

ROC Curve

Various classification algorithms can be configured to produce a class label based on a predefined threshold on the probability of the data item belonging to a class. For example, in the following chart, we can see that the classifier's predictions can strongly differ based on the threshold. This indirectly affects the precision and recall, sensitivity and specificity.

Sensitivity, or recall, as we saw in the previous section, is the ratio of True Positives and total positive data items. It is also called True Positive Rates. Specificity is the ratio of True Negatives and all the data items that are actually negative. Usually, there is a trade-off between the two.

In the following experiments, we will train a logistic regression model and find True Positive Rate and False Positive Rate based on classification output for different threshold of the model. Let's use Scikit-learn's inbuilt functionality to generate synthetic dataset for such experiments:

```
from sklearn.datasets import make_moons
X1, Y1 = make_moons(n_samples=1000, shuffle=True, noise=0.1)
```

Let's have a look at the dataset we thus generated using matplotlib. The results should appear similar to Figure 8-3.

```
import matplotlib.pyplot as plt
plt.figure(figsize=(8, 8))
plt.scatter(X1[:, 0], X1[:, 1], marker='o', c=Y1, s=25, edgecolor='k')
plt.show()
```

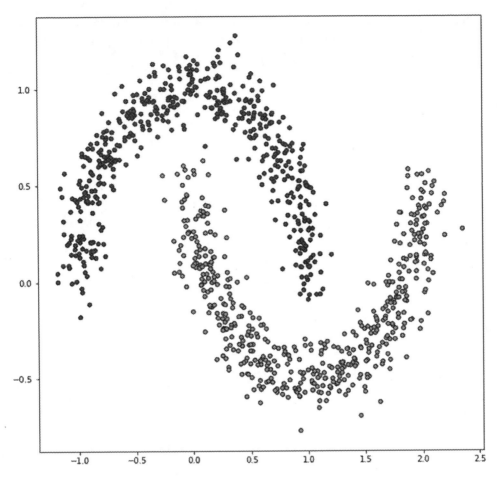

Figure 8-3. *Displaying randomly generated dataset*

For generating less complex dataset, you can use

```
from sklearn.datasets import make_classification
X1, y1 = make_classification(n_classes=2, n_features=2, n_redundant=0,
n_informative=1, n_clusters_per_class=1)
```

We will train a logistic regression model.

```
from sklearn.linear_model import LogisticRegression
from sklearn.model_selection import train_test_split
X_train,X_test,y_train,y_test = train_test_split(X1,y1,test_
size=0.2,random_state=42)
logreg = LogisticRegression()
logreg.fit(X_train,y_train)
```

Instead of predict() method, we can call predict_proba() method that produces each point belonging to the second class (class label=1).

```
logreg.predict_proba(X_test)
```

This will give you an array of shape (20,2). There are 20 test samples, for which the probability of each class is given in the two columns. We will take one of the columns for the probability and manipulate the threshold and monitor the effect it has on TPR and FPR.

```
y_pred_proba = logreg.predict_proba(X_test)[:,1]
from sklearn.metrics import roc_curve
[fpr, tpr, thr] = roc_curve(y_test, y_pred_proba)
```

The objects returned from the last statement can be used to analyze the effect of setting different thresholds. The curve can be traces across multiple values of threshold. Before that, we will introduce one more metric, which takes the false positive rates and true positive rates and, based on them, computes the area under the curve thus generated.

```
from sklearn.metrics import auc
auc (fpr, tpr)
>> 0.8728442728442728
```

Let's plot the graph.

```
plt.figure()
plt.plot(fpr, tpr, color='coral', label = 'ROC Curve with Area Under Curve = '+str(auc (fpr, tpr)))
plt.xlabel('False positive Rate (1 - specificity)')
plt.ylabel('True Positive Rate ')
plt.legend(loc='lower right')
plt.show()
```

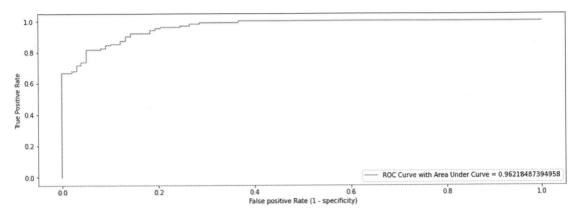

Figure 8-4. *Plotting ROC curve based on False Positive Rate and True Positive Rate*

Figure 8-4 shows the actual characteristics of the quality of prediction. The area under the curve should be as high as possible. ROC curve provides a more standardized way to compare multiple models regardless of the threshold which might affect the end results.

Overfitting and Regularization

We can fine-tune the models to fit the training data very well. In this process, we often play with several properties of the algorithms that may directly manipulate the complexity of the models. Let's try to play with linear regression and use a more complex model to fit the training data points from the last chapter more precisely. We will create a new set of features to take simple arithmetic transformation of the independent variable and fit linear regression based on them. This method is called polynomial regression.

Note   In polynomial regression, we are still using the linear regression methodologies to fit a line. However, this is done after expanding the independent variables into polynomial features.

In the following code block, we will use the same students' marks-salary dataset we used in the previous chapter. We will expand the features:

```
import numpy as np
import pandas as pd
```

```
from sklearn.linear_model import LinearRegression
from sklearn.preprocessing import PolynomialFeatures

data = pd.DataFrame({"marks":[34,51,64,88,95,99], "salary":[3400, 2900,
4250, 5000, 5100, 5600]})


X = data[['marks']].values
y = data['salary'].values

poly = PolynomialFeatures(3)
X1 = poly.fit_transform(X)
```

At this point, we have a new array X1 of shape (6,4), which are created from x^0, x^1, x^2, and x^3. This should look like

```
array([[1.00000e+00, 3.40000e+01, 1.15600e+03, 3.93040e+04],
       [1.00000e+00, 5.10000e+01, 2.60100e+03, 1.32651e+05],
       [1.00000e+00, 6.40000e+01, 4.09600e+03, 2.62144e+05],
       [1.00000e+00, 8.80000e+01, 7.74400e+03, 6.81472e+05],
       [1.00000e+00, 9.50000e+01, 9.02500e+03, 8.57375e+05],
       [1.00000e+00, 9.90000e+01, 9.80100e+03, 9.70299e+05]])
```

We will use LinearRegression the same way we did before. However, to find the regression line (or lines), we will prepare an array containing marks from minimum marks in the dataset to maximum marks in the dataset, transform it to the same polynomial features, and predict the values. This time, we will learn four parameters instead of just two. The plot we thus create will show the regression line.

```
reg = LinearRegression()
reg.fit(X1, y)

X_seq = np.linspace(X.min(),X.max(),100).reshape(-1,1)
X_seq_1 = poly.fit_transform(X_seq)
y_seq = reg.predict(X_seq_1)

import matplotlib.pyplot as plt
plt.figure()
plt.scatter(X,y)
plt.plot(X_seq, y_seq,color="black")
plt.show()
```

This will print the graph as shown in Figure 8-5 that shows the predicted values for each possible point in our range.

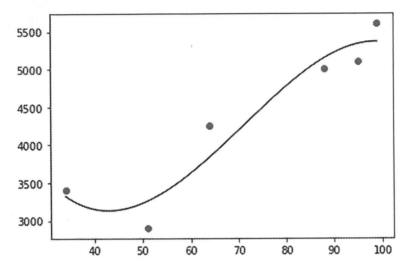

Figure 8-5. *Linear regression with polynomial features*

You can see that we can increase the complexity and the predictions fit the training data more closely, thus reducing the overall error and improving the accuracy. This looks promising but might be misleading.

Remember while solving any machine learning problem, you have to choose a function to fit the training data from among a set of hypotheses. Usually, the size of data is very low compared to the possible data your system might need to predict once it is deployed in the real world. Our efficiency, accuracy, as well as the measure of accuracy are all limited by the quality and the quantity of labelled data that we are able to generate, collect, or annotate. In the preceding regression example, we tried to fit a third-degree polynomial curve in order to minimize the residuals based on actual dependent value. Though there is a general uptrend in the data, the model might be misleading.

The following modification allows you to find and plot predicted values for marks from 0 to 100 instead of minimum and maximum of the training data.

```
X_seq = np.linspace(0,100,100).reshape(-1,1)
X_seq_2 = poly.fit_transform(X_seq)
y_seq = reg.predict(X_seq_2)

plt.figure()
plt.scatter(X,y)
```

```
plt.plot(X_seq, y_seq,color="black")
plt.show()
```

The curve thus generated looks like the following.

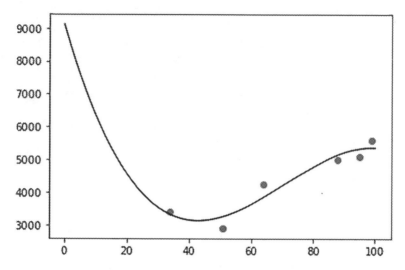

Figure 8-6. *Overfitting curve because of overcomplex regression model*

Here in Figure 8-6, we tried to increase the complexity of the model so that it captures the training data better; however, this led to unexpected errors for the data that wasn't present in the real data. You can see that due to the leftmost point, the predictions for marks less than that show higher salaries for lower marks. The issue is that we made the model more complex to fit the data too well. This is called **overfitting**.

Opposite of this situation is a model that doesn't learn from the training data enough. If we have one independent variable, we learn two parameters – y-intercept and the slope of the regression line. In the preceding example, we increased the number of parameters to fit the line much better.

If we instead reduce the number of parameters, say, 1, we are drastically reducing the complexity of the model, and we will be able to capture less details from the training data. In our example, the model with only one parameter will return the average salary based on the training data, and thus, a horizontal line parallel to the x axis will represent the predictions as shown in Figure 8-7. That is, the model will predict the same salary regardless of the marks a student attains. This is called **underfitting**.

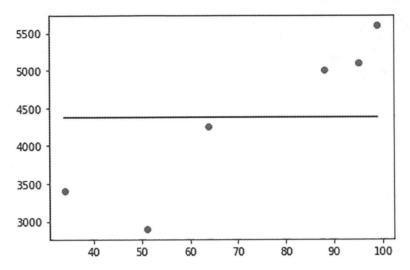

Figure 8-7. *Underfitting curve due to overly simplistic regression model*

Bias and Variance

Bias and variance are the properties of a model that arise due to either oversimplicity or overcomplexity of our model. Bias, in general, represents how far a model's predictions are compared to the actual values. A model with high bias means that the model is overly simple, and the assumptions it has learned are too basic. For that reason, the model isn't able to properly capture the necessary patterns in the data. Thus, the model has high error while training, as well as while predicting.

Variance represents how sensitive the model is to fluctuations in the data. Say, we have a data point that represents a student who obtained 35 marks and a salary of $6000 and another data point for a student who obtained 34 marks and a salary of $2000, and the system tries to learn the difference from both; this can cause huge difference in how the predictions are generated. When the variance is high, the model will capture all the features of the dataset, including the noise and randomness. Thus, it becomes overly tuned. However, when it encounters unseen data, it might yield unexpectedly poor results. Such a model yields a low training error; however, the error is quite high while testing.

We need to find a balance between bias and variance in order to come up with a model that is sensitive to patterns in our data while also being able to generalize to new unseen data. The trends in error with respect to model complexity are shown in Figure 8-8.

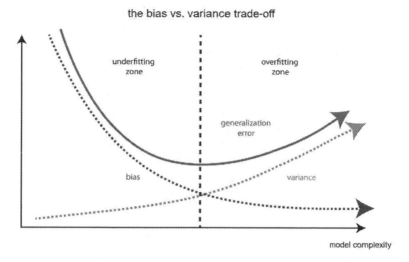

Figure 8-8. *Bias-variance trade-off*

Regularization

A way is to manipulate the cost function that penalizes overcomplexity in the model in order to find the right parameters instead of explicitly limiting the numbers of parameters to learn. Thus, say, the cost function of an algorithm is given by J(w), where w represents the weights of the parameters; the new cost function is given by

$$J(w) = J_D(w) + \lambda J(w)$$

where the first part is the sum of squared residuals, as we've seen in the previous section, and the additional part is λ times sum of all the weights. λ is a hyperparameter that is used to control how much regularization strength we want to apply. Due to the additional part in the cost function, the parameter weights at any instance contribute directly to the cost, which the algorithm tries to minimize. Due to this, if any of the weights are too high, the cost increases and the algorithm tries to find a step away from those sets of weights.

Eventually, we reach a middle ground where the model is complex enough to capture the essence of the structure of the training data while penalizing over complexity so that we avoid learning from extreme outliers and noise.

L1 and L2 Regularization

Regularization is a technique that discourages learning a more complex or flexible model, so as to avoid the risk of overfitting by manipulating the cost function to avoid learning weights that are too high. L2 regularization is a poly-time closed-form solution that penalizes the model based on the summation of squares of model weights. It helps reduce the overfitting but doesn't produce a sparse solution. In Ridge Regression, L2 penalty is applied, which modified the cost function to

$$J(w) = \sum_{i=1}^{n}\left(y_i - w^T x_i\right)^2 + \lambda \sum_{j=0}^{m} w_j^2$$

The coefficient estimates produced by this method are also known as the L2 norm.

In Lasso Regression, L1 penalty is applied. It uses the absolute value of the weights (instead of the squares as in L2 penalty). The cost function thus becomes

$$J(w) = \sum_{i=1}^{n}\left(y_i - w^T x_i\right)^2 + \lambda \sum_{j=0}^{m} |w|$$

In many experiments, we will see that with logistic regression and linear regression, L1 regularization leads to a lot of model weights converging to zeros (0). This implies that the model with regularization has learned that the impact of some of the features can be considered negligible in order to achieve a solution that is not overfitted.

Let's do a simple experiment to find the impact of these two methods on the parameters.

```
import numpy as np
import pandas as pd
from sklearn.linear_model import LinearRegression, Lasso, Ridge
from sklearn.preprocessing import PolynomialFeatures
import matplotlib.pyplot as plt

data = pd.DataFrame({"marks":[34,51,64,88,95,99], "salary":[3400, 2900,
4250, 5000, 5100, 5600]})

# data.bill
X = data[['marks']].values
y = data['salary'].values
```

```
fig, axs = plt.subplots(1,3, figsize=(15,5))
methods = ['Polynomial Regression', 'Lasso Regression alpha=1', 'Ridge
Regression alpha=1']

for i in [0,1,2]:
    poly = PolynomialFeatures(3)
    X1 = poly.fit_transform(X)
    if i==0:
        reg = LinearRegression()
        reg.fit(X1, y)
    if i==1:
        reg = Lasso(alpha=1)
        reg.fit(X1, y)
    if i==2:
        reg = Ridge(alpha=1)
        reg.fit(X1, y)

    X_seq = np.linspace(0,X.max(),100).reshape(-1,1)
    X_seq_1 = poly.fit_transform(X_seq)
    y_seq = reg.predict(X_seq_1)

    axs[i].scatter(X,y)
    axs[i].plot(X_seq, y_seq,color="black")
    axs[i].set_title(methods[i])
plt.show()
```

In this code example, we reattempt the problem of predicting students' salary based on the marks they obtained. Here, we have first converted the only independent variable (marks) into polynomial features and then trained three models based on that. The first one is regression without regularization, the second one is Lasso Regression, and the third one is Ridge Regression. The effect of the two techniques is evident from the regression line they produce as shown in Figure 8-9.

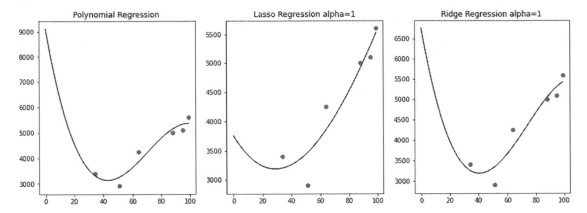

Figure 8-9. *Visualizing effects of Lasso Regression and Ridge Regression in our use case*

Thus, we see that regularization reduces the variance in the models significantly without substantial increase in bias. It is important to tune the algorithm by experimenting with the values of regularization strength λ to make sure that we don't lose any important patterns in the data.

Hyperparameter Tuning

While approaching a machine learning problem, you have to engineer and select the right features, pick the algorithm, and tune the selected algorithm (or algorithms) for the hyperparameters they are affected by.

Note   The terms hyperparameters and parameters can't be used interchangeably. Parameters are the weights a model learns during learning phase. Hyperparameters are the externally controlled elements that affect how and what the model learns.

You might often face the choices like the following:

- "K" in K-nearest neighbors

- Regularization strength in Ridge Regression and Lasso Regression

- Maximum depth of a decision tree

- Learning rate for gradient descent

Effect of Hyperparameters

We will do a simple experiment to see how closely we fit a synthetic dataset using the first two columns based on the hyperparameters we can tune for logistic regression.

Let's create a dataset using Scikit-learn's make classification functionality.

```
from sklearn.datasets import make_classification

X, y = make_classification(n_samples=400, n_features=2, n_informative=2, n_redundant=0)
```

We will need to create separate training dataset and test dataset for analyzing the accuracy.

```
X_train, X_test, y_train, y_test = train_test_split(X,y, test_size=0.3)
```

Now we will use logistic regression with polynomial features. We will iteratively try different polynomial degree and see the difference it makes in the accuracy.

```
accuracy_history = []
for i in range(1,15):
    poly = PolynomialFeatures(i)
    X1 = poly.fit_transform(X_train)
    reg = LogisticRegression(max_iter=100)
    reg.fit(X1, y_train)
    X1_test = poly.transform(X_test)
    y_pred = reg.predict(X1_test)
    accuracy_history.append(accuracy_score(y_test, y_pred))
```

Let's plot the accuracy:

```
import matplotlib.pyplot as plt
plt.plot(accuracy_history)
```

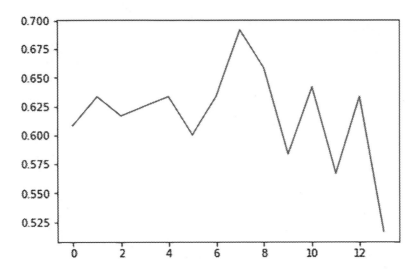

Figure 8-10. *Plotting accuracy with respect to polynomial features*

We can see in Figure 8-10 that the accuracy increases from degree 5 to degree 7 and then decreases again. Based on this analysis with only one hyperparameter, which is the degree of the polynomial, find the hyperparameter value that yields the best accuracy. In most practical scenarios, you should prefer cross validation instead of working with dedicated training and testing dataset.

If we have multiple hyperparameters, we have to evaluate the model for multiple possible values for each hyperparameter. Let's look at the scenario for creating decision tree classifier. These are some of the hyperparameters you can tune:

- criterion: Either Gini based or entropy based.

- max_depth: Maximum depth of a tree.

- min_samples_split: Minimum number of samples that are required to split a node. It can be either an integer representing the number or a float that represents a fraction of total samples.

- min_samples_leaf: Minimum number of samples that should be present in both right leaf and left leaf.

We discussed the working of decision tree in the previous chapter. You can see that such decisions can impact the quality of end results. Let's say we have the following possibilities:

- criterion: Gini, entropy (two possible values)

- max_depth: None, 5, 10, 20 (four possible values)

- min_samples_split: 4, 8, 16 (three possible values)

- min_samples_leaf: 1, 2, 4 (three possible values)

By considering all the options, we will build up to 2x4x3x3 = 72 decision trees, out of which, we will select the one that gives the best metrics. It is possible to make the selections with the use of multiple loops – but Scikit-learn provides ready-to-use implementations to exhaust all the possibilities (grid search) and test random possibilities (random search).

Grid Search

Grid search, or parameter sweep, is the process of searching through a specified subset of hyperparameter spaces exhaustively. For the example given previously, some of the 72 hyperparameter combinations will be

1. {criterion:gini, max_depth:None, min_samples_split:4, min_samples_leaf:1}

2. {criterion:gini, max_depth:None, min_samples_split:4, min_samples_leaf:2}

3. {criterion:gini, max_depth:None, min_samples_split:4, min_samples_leaf:4}

4. {criterion:gini, max_depth:None, min_samples_split:8, min_samples_leaf:1}

5. {criterion:gini, max_depth:None, min_samples_split:8, min_samples_leaf:2}

Let's initialize the DecisionTreeClassifier as base classifier and prepare the list of possible parameter values.

```
from sklearn.tree import DecisionTreeClassifier
from sklearn.model_selection import GridSearchCV

param_grid= {"criterion":["gini", "entropy"], "max_depth":[None,5,10,20],
"min_samples_split":[4,8,16], "min_samples_leaf":[1,2,4]}
base_estimator = DecisionTreeClassifier()
grid_search_cv = GridSearchCV(base_estimator, param_grid, verbose=1, cv=3)
```

We will prepare another synthetic dataset. This prepares points that can be visualized as two half-circular figures.

```
from sklearn.datasets import make_moons
from sklearn.model_selection import train_test_split
dataset= make_moons(n_samples=10000, shuffle=True, noise=0.4)
X_train,X_test,y_train,y_test = train_test_split(dataset[0],dataset[1],te
st_size=0.2,random_state=42)
```

Rather than preparing the loops ourselves with possible values, we will use the fit() method of grid_search_cv that will prepare the 72 models and collect their results. We have already initialized a GridSearchCV object with our parameter grids and threefold cross validation. We have marked verbose=1, which will make the fit() method print the progress details.

```
grid_search_cv.fit(X_train, y_train)
```

This should print

```
Fitting 3 folds for each of 72 candidates, totalling 216 fits
```

This confirms that we have prepared 72 models internally, which, combined with threefold search, generated a total of 216 individual decision trees. All the details are returned in the grid_search_cv.cv_results_ object. You can look at all the information you can extract using

```
grid_search_cv.cv_results_.keys()
```

which should return

```
dict_keys(['mean_fit_time', 'std_fit_time', 'mean_score_time', 'std_score_
time', 'param_criterion', 'param_max_depth', 'param_min_samples_leaf',
'param_min_samples_split', 'params', 'split0_test_score', 'split1_test_
score', 'split2_test_score', 'mean_test_score', 'std_test_score', 'rank_
test_score'])
```

To simplify the interpretation, you can directly find the best accuracy using grid_search_cv.best_score_. This is the score associated with the best estimator that can be returned using

```
grid_search_cv.best_estimator_
```

which returns

```
DecisionTreeClassifier(criterion='entropy', max_depth=5, min_samples_
leaf=4, min_samples_split=4)
```

You can get all the additional parameters using

```
grid_search_cv.best_estimator_.get_params()
```

Random Search

Rather than exhaustively searching for all the combinations in the parameter space, random search selects random possibilities and selects the best model accordingly. We can provide a distribution instead of discrete values to define a search space.

You can initialize the parameter grid like the following:

```
from scipy.stats import randint
from sklearn.model_selection import RandomizedSearchCV
from sklearn.tree import DecisionTreeClassifier
base_estimator = DecisionTreeClassifier()
param_grid= {"criterion":["gini", "entropy"], "max_depth":randint(1, 20),
"min_samples_split": [1,2,4]}
random_search_cv = RandomizedSearchCV(estimator = base_estimator, param_
distributions = param_grid, n_iter = 100, cv = 5, verbose=2)
random_search_cv.fit(X_train, y_train)
```

Here, we have initialized `parameter_grid` to pick one of the two criteria, randomly generated `max_depth`, and one of the three options for `min_samples_split`. For real values, you can also use `scipy.stats.norm as the value of relevant param_ grid key.`

One difference here is that you have to specify the `n_iter` value that mentions the number of individual estimators that will be created based on a random combination of hyperparameter values we have defined.

In practice, the number of estimators generated while using grid search will increase the search space as the number of hyperparameters increases, which might lead to some situations that might not be practical to work with. However, due to exhaustive nature of grid search, it will find the best possible set of parameters.

In such cases, it is a good idea to begin experimenting with random search, which will help you reduce the search space. You can then perform a thorough grid search with a limited set of hyperparameters and possible values to find the best model thoroughly.

Summary

This chapter has provided us with the essentials that will be used in any machine learning experiment to evaluate and tune the models. In the next chapter, we will study more supervised learning methods.

## CHAPTER 9

Supervised Learning Methods: Part 2

We have discussed several regression and classification algorithms and found out ways on efficiently validating the performance of the models and tuning them well. We will extend our repository of tools and techniques through some more popular algorithms that have proven to be state of the art for various machine learning problems in the past. We will visualize the decision boundaries of the classification algorithms and their variations to understand the mechanism and the results more practically.

Naive Bayes

Naive Bayes is a category of Bayesian classifiers that aim to predict the probability that a given data point belongs to a particular class. These methods are based on Bayes' theorem, which is one of the fundamental theorems in probabilities.

Naive Bayes classifiers have a general assumption[1] that the effect of an attribute value on a given class in independent of the values of the other attributes. This assumption is called class-conditional independence. It simplifies the computations involved and thus considered naive.

---

[1] Han, Jiawei, et al. Data Mining: Concepts and Techniques, Third Edition. 3rd ed. Morgan Kaufmann Publishers, 2012

149

© Ashwin Pajankar and Aditya Joshi 2022
A. Pajankar and A. Joshi, *Hands-on Machine Learning with Python*, https://doi.org/10.1007/978-1-4842-7921-2_9

Bayes Theorem

Bayes theorem is a formula to compute conditional probability, named after Thomas Bayes, who was a clergyman in the 18th century. The theorem computes the probability of an event based on prior knowledge of conditions that might be related to the event. It is expressed as

$$P(A|B) = \frac{P(B|A)P(A)}{P(B)}$$

This provides the probability of A happening, given that the event B has happened. Consider a prediction example.

You are developing a detection system that captures blips in the radar. Say, the event A represents an enemy being present in the space. The event B represents a blip in the radar.

Remember, a blip in the radar doesn't always mean that the enemy is present. Similarly, it is also possible that the enemy might be present in the area but no blip is present in the radar. Thus, the formula given previously will help us find the probability of an enemy being present in the area given a blip has occurred in the radar.

To compute this probability, we need to collect a lot of the past data. Past data can directly provide us the probability of enemy being in the space, P(A), and probability of seeing a blip in the radar, P(B). These probabilities are respectively called prior probability and marginal probability.

A thorough analysis of past events would also help us find the likelihood, or the probability of seeing a blip in the radar, given that we have the prior knowledge that an enemy is in the area. This likelihood is computed by P(B|A), which, in our machine learning experiments, will be extracted from the training data.

Conditional Probability

The main component of the formula given by Bayes' theorem is conditional probability. Conditional probability is a measure of the probability of an event, given that another event (by assumption, presumption, assertion, or evidence) has already occurred. It is computed from joint probability of A and B.

$$P(A|B) = \frac{P(A \cap B)}{P(B)}$$

Here, P(A∩B) is the probability that both A and B occurred.

If we modify this formula for P(B|A) and try to match the transposed terms for $P(A \cap B)$, we will end up with Bayes' theorem.

How Naive Bayes Works

In simple words, during the training phase, naive Bayes attempts to calculate the prior probability for the given class labels. This is followed by computing the likelihood probability with each attribute for each class. This information is used to compute the posterior probability using the Bayes formula as we've seen in the previous section. We calculate the posterior probability for the data item belonging to each class and choose the class label with the highest probability.

In many practical cases, we assume that the data follows Gaussian distribution, due to which the probability of an item, given a class label c, can be defined as

$$P\left(X|Y=c\right)=\frac{1}{\sqrt{2\pi\sigma_c^2}}e^{\frac{-(c-\mu_c)^2}{2\sigma_c^2}}$$

where μ and σ are the mean and variance of the continuous data point attributes for a given class c. This is especially valid when you have continuous data. If the data doesn't follow normal distribution, it might be possible to transform it first.

It is possible that there are no training samples for one of the classes, which might lead to P(X|Y=c) as 0 (zero) for some cases. Plugging this value in the Bayesian equation will lead to a zero probability, which might be misleading in some cases. It also cancels the effect of other probabilities involved in the final computation. This is resolved by adding 1 (one) to each count before computing the probabilities. The error added by doing this is negligible for sufficiently large datasets, but the positive effect is impactful. This is called as Laplacian correction.

Multinomial Naive Bayes

In some cases, instead of Gaussian distribution, we assume that the data follows multinomial distribution due to the presence of discrete counts instead of continuous data. This is the case when we have k possible mutually exclusive outcomes, and

151

we study the number of times outcome *i* over the *n* trials, X=(X$_1$, X$_2$, ... X$_k$) follow a multinomial distribution with probabilities p = (p$_1$, p$_2$, ... p$_k$) denote the probabilities of each possible outcome. The probability parameter would change to

$$\theta_{yi} = \frac{N_{yi} + \alpha}{N_y + \alpha n},$$

where N$_{yi}$ is the number of times a feature I appears in a sample of class y in the training data and N$_y$ is the total count of all features for class y. α is a parameter that controls smoothing. Multinomial naive Bayes has proven to be highly effective in short text classification, especially examples like sentiment analysis on short social media (twitter) text.

Naive Bayes in Python

For this example, we will load Iris dataset from the available datasets. Because Iris dataset contains continuous variables, we will assume Gaussian distribution and train a Gaussian naive Bayes model.

```
from sklearn import datasets
import matplotlib.pyplot as plt
iris = datasets.load_iris()
X = iris.data[:, :]  # we can select individual features
y = iris.target
```

Remember that naive Bayes inherently captures the probabilities with respect to multiple classes. So unlike a few other methods, techniques like One-vs-Rest are not required here. We will not head to split the dataset into train and test set while holding 25% of the data for testing.

```
from sklearn.model_selection import train_test_split
X_train, X_test, y_train, y_test = train_test_split(X, y, test_size = 0.25,
random_state = 0)
```

We will now import Gaussian naive Bayes from sklearn and fit it to find the parameters and probabilities.

```
from sklearn.naive_bayes import GaussianNB
clf = GaussianNB()
clf.fit(X_train, y_train)
```

You can explore the learned parameters using

```
clf.class_prior_  # prior probabilities for the three classes
clf.sigma_ # variance of each feature, with respect to each class
clf.theta_ # mean of each feature, with respect to each class
```

As we've discussed in the previous section, if we know the prior probabilities and mean and variance of each feature, we can use it to compute the conditional probability $P(X|Y = c)$, which can in turn be used to predict $P(Y=c | X)$ using Bayes' theorem. Prediction method follows the same conventions:

```
y_pred = clf.predict(X_test)
from sklearn.metrics import confusion_matrix
cm = confusion_matrix(y_test, y_pred)
array([[13,  0,  0],
       [ 0, 16,  0],
       [ 0,  0,  9]], dtype=int64)
```

The dataset is simple and small, and the classifier has learned the right distribution to predict the unseen data correctly. To be able to better appreciate the model, we can visualize the decision boundaries. In the next piece of code, we will retrain the classifier with only the first two columns from the data so that we can visualize the decision boundaries in two dimensions.

```
from matplotlib.colors import ListedColormap
import numpy as np

clf = GaussianNB()
clf.fit(X_train[:,:2], y_train)
```

Here, we trained a new model with only the first two columns of X_train. This will give us a clearer view of what the classifier is learning.

```
x_set, y_set = X_train[:,:2], y_train
X1, X2 = np.meshgrid(np.arange(start = x_set[:, 0].min() - 1, stop =
x_set[:, 0].max() + 1, step = 0.01),
                      np.arange(start = x_set[:, 1].min() - 1, stop =
                      x_set[:, 1].max() + 1, step = 0.01))
```

```
plt.contourf(X1, X2, clf.predict(np.array([X1.ravel(), X2.ravel()]).T).
reshape(X1.shape),
            alpha = 0.75, cmap = ListedColormap(('purple', 'green',
            'yellow')))
plt.xlim(X1.min(), X1.max())
plt.ylim(X2.min(), X2.max())
for i, j in enumerate(np.unique(y_set)):
    plt.scatter(x_set[y_set == j, 0], x_set[y_set == j, 1],
                color = ListedColormap(('purple', 'green', 'yellow'))(i),
                label = j)
plt.legend()
plt.show()
```

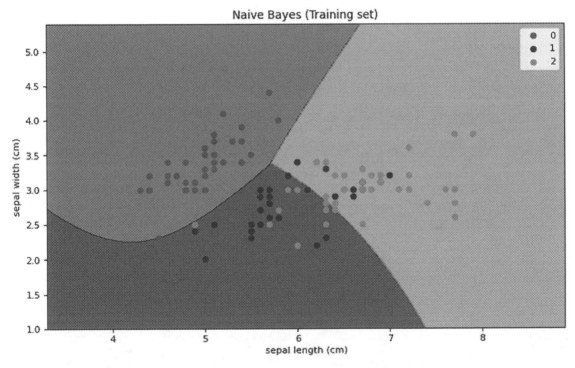

Figure 9-1. *Decision boundary of naive Bayes classifier trained on Iris dataset*

Figure 9-1 shows the decision boundary plotted through the code explained in the previous page. Though we are displaying the data in first two dimensions, the overlap between the second two classes looks more prominent than is actually is – but this visualization shows the flexibility and robustness of Bayesian methods in learning the decision boundaries.

154

Support Vector Machines

Support Vector Machines (SVM) is a highly efficient and effective method that has been proven to be a state of the art in various scenarios. It is used for classification of both linear and nonlinear data.

Unlike other classification methods that attempt to find a decision boundary that minimizes the error between two classes, Support Vector Machines try to find a decision boundary that maximizes the margin between the two classes called a maximum marginal hyperplane. For that reason, Support Vector Classifiers are also called Maximum Margin Classifiers.

In Figure 9-2, you can see that there are multiple possible decision boundaries between the two classes. However, Support Vector Machines attempt to find the boundary that maximizes the separation between the two classes.

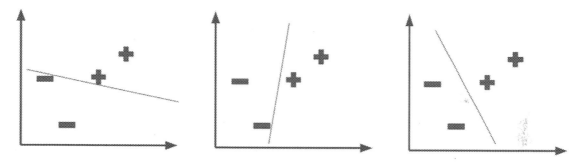

Figure 9-2. *Possible decision boundaries between two classes*

The margin around the decision boundary depends on the points that generate the maximum width of the margin and thus the difference between the closest training data points. These points are called **support vectors**, and thus the name of the method. The other points in the training dataset do not contribute to the model. While learning, the goal is to maximize the width of the margin and minimize the margin errors, that is, the points that are on the wrong side of the margin.

How SVM Works

The hyperplane that divides the data points has equal margin on the two sides parallel to the hyperplane. In two dimensions, hyperplane is a straight line that divides the data into positive class and negative class. Figure 9-3 shows the margins on the two sides of the line running parallel and equidistant to the hyperplane.

The hyperplane can be written as

$$W \cdot X + b = 0$$

Here, W is the weight vector, and b is the bias (scaler). In two dimensions, A point (x_1, x_2) can be considered to be on the hyperplane. We can instead write this as

$$w_1 x_1 + w_2 x_2 + b = 0$$

Or the bias b can be expressed as an additional weight w_0 to simplify the implementation:

$$w_0 + w_1 x_1 + w_2 x_2 = 0$$

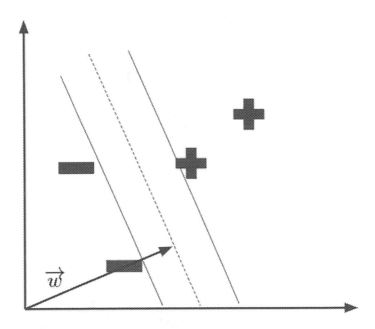

Figure 9-3. *Maximum margin hyperplace*

For any point above the separating hyperplane, belonging to the positive class y = +1,

$$w_0 + w_1 x_1 + w_2 x_2 \geq 0 \ for \ y = +1$$

And for any point below the separating hyperplane, belonging to the negative class y = -1,

$$w_0 + w_1 x_1 + w_2 x_2 \leq 0 \ for \ y = -1$$

We can combine these two equations and write the same in a simplified form as

$$y_i \left(w_{i0} + w_{i1}x_{i1} + w_{i2}x_{i2} \right) \geq +1$$

which is true for all points (x_{i1}, x_{i2}), belonging to class y_i, that are located on the margins. These are called support vectors.

As W is the weight vector comprising $\{w_1, w_2\}$, we can write the distance of the margin from the decision boundary as $\dfrac{1}{\|W\|}$ w where $\|W\|$ is the Euclidean norm of W. The distance from other margin will also be the same. Thus, the distance between the two margins is $\dfrac{2}{\|W\|}$. The aim of the learning algorithm is to find the support vectors and the maximum margin hyperplane, that is, the decision boundary that maximizes this distance. This resolves to a constrained quadratic optimization problem, which is solved using a Lagrangian formulation. Explanation of this solution is out of the scope of this book.

Beyond this mathematical optimization, SVMs use another way to find a decision boundary in case of data items belonging to different classes that are not linearly separable.

Nonlinear Classification

As SVM can create a linear decision boundary, it is important to understand how they can be customized to find decision boundaries in nonlinearly separable data. Here's an intuition of utilizing nonlinearly separable data belonging to the two classes in only one dimension for simplicity. The separating hyperplane here would be a point (anything beyond a certain point belongs to the positive class). But due to the nature of the spread of data, it is not possible to find one such point.

Using a simple transformation, we add one more dimension to the data – and as shown in Figure 9-4, it is possible to find a hyperplane (now, a line) that can easily separate the two classes.

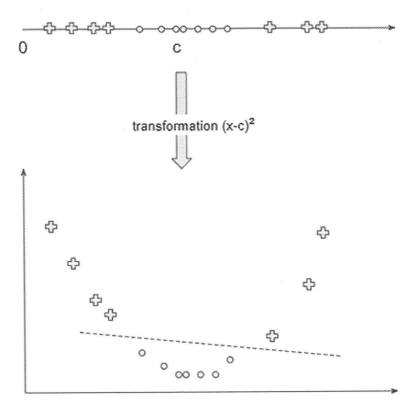

Figure 9-4. *Transformation of original data from 1D to 2D leads to a straight decision boundary*

So the idea is first transform the data to a higher dimension using a nonlinear transformation. Then we can find a hyperplane in the new dimensions that can easily separate the two classes. At this point, we can use the same method for SVM as mentioned in the previous section. However, it is inefficient to apply plenty of such transformations, followed by dot products. This is solved using a method called the Kernel Trick.

Kernel Trick in SVM

In the quadratic optimization, we require to find the linear transformation only while finding the dot products. Say, a data point X_i can be transformed to the new space using $\varphi(X_i)$, the dot product that is often required will be $\varphi(X_i) \bullet \varphi(X_j)$ for all the support vectors X_i and Xj. A Kernel is a mathematical function that finds the dot product in the transformed space. Thus, we can write

$$K\left(X_i, X_j\right) = \varphi\left(X_i\right) \cdot \varphi\left(X_j\right)$$

So during the training process, if there's any situation where we first need the transformation, followed by dot products, we can replace that with the Kernel function. The computations happen in the lower dimensions, and the decision boundary can be seen as being created in the higher dimensions. This is known as Kernel Trick. The most commonly used Kernels are given here:

Polynomial Kernel

$$k(\mathbf{x_i}, \mathbf{x_j}) = (\mathbf{x_i} \cdot \mathbf{x_j} + 1)^d$$

Gaussian Kernel

$$k(x, y) = \exp\left(-\frac{\|x - y\|^2}{2\sigma^2}\right)$$

Gaussian Radial Basis Function (RBF)

$$k(\mathbf{x_i}, \mathbf{x_j}) = \exp(-\gamma\|\mathbf{x_i} - \mathbf{x_j}\|^2)$$

$$\gamma = 1/2\sigma^2$$

Laplace RBF kernel

$$k(x, y) = \exp\left(-\frac{\|x - y\|}{\sigma}\right)$$

Hyperbolic tangent kernel

$$k(\mathbf{x_i}, \mathbf{x_j}) = \tanh(\kappa \mathbf{x_i} \cdot \mathbf{x_j} + c)$$

In SVM implementations, you can experiment with the different possible kernels through the hyperparameters, which you will have to tune in the experimental process while also keeping a note of the additional hyperparameters that your Kernel of choice depends on. For example, if you choose the polynomial kernel in Scikit-learn, you will define kernel as poly and add a degree parameter to decide the degree of polynomial you want to use.

Support Vector Machines in Python

We will experiment with a simple dataset and visualize the decision boundaries with different kernels of SVM implementation. Scikit-learn provides Support Vector Machines under `sklearn.svm` package. SVM is a class of algorithms that can be used in regression problems, often written as SVR (Support Vector Regressors), and classification problems, written as SVC (Support Vector Classifiers).

sklearn.svm.LinearSVC is implemented using liblinear and offers a different set of penalties and loss functions compared to sklearn.svm.SVC, which is based on libsvm. For large datasets, LinearSVC has been recommended instead of SVC. We will compare LinearSVC and SVC with different types of kernels in the following examples.

Choosing the right kernel requires experimentation through cross validation–based model selection. Linear and RBF have proven to be fast and highly effective in a lot of common use cases.

We will continue this example with Iris dataset while taking first two columns only: first, to be able to visualize well and, second, to make decision boundaries only based on limited parameters.

```
import numpy as np
import pandas as pd
from sklearn.datasets import load_iris

iris = load_iris()
df = pd.DataFrame(iris.data, columns=iris.feature_names)
df['species'] = pd.Categorical.from_codes(iris.target, iris.target_names)
df['target'] = iris.target
X = df.iloc[:,:2]
y = df.iloc[:,5]
```

We can train an SVM with default hyperparameters as follows:

```
svc = svm.SVC(kernel='linear', C=C).fit(X, y)
```

SVC can accept these kernels: linear, poly, rbf, sigmoid, and precomputed. The default value is rbf, or Radial Basis Function. Remember, you might need to supply additional hyperparameters depending on your kernel of choice.

For example, if you decide to build a Support Vector Classifier with polynomial kernel, you may add another hyperparameter for degree. If you do not provide the additional hyperparameter, the default value will be taken, which is 3.

Once the model is trained, you can look at the support vectors – which are representatives of the dataset boundaries.

```
support_vector_indices = svc.support_
print(len(support_vector_indices))
```

```
support_vectors_per_class = svc.n_support_
print(support_vectors_per_class)
70
[ 2 34 34]
```

These properties give the count of support vectors. In the second output, you can see that there are only two support vectors for the first class; that is, Iris Setosa, which we've seen in previous chapters, is relatively easier to separate.

You can find the actual points acting as support vectors:

```
support_vectors = svc.support_vectors_
```

support_vectors object will contain every training data point that is being used in constructing the decision boundaries that maximizes the hyperplane. You can visualize them to get a clearer understanding of what's being learned. The following code would lead to results similar to Figure 9-5.

```
xx,yy = np.meshgrid( np.arange(x_min, x_max, 0.1), np.arange(y_min,
y_max, 0.1)  )
Z = clf.predict(np.c_[xx.ravel(), yy.ravel()])
Z = Z.reshape(xx.shape)

# Visualize support vectors
plt.scatter(X.iloc[:,0], X.iloc[:,1])
plt.contourf(xx, yy, Z, cmap=plt.cm.coolwarm, alpha=0.4)
plt.scatter(support_vectors[:,0], support_vectors[:,1], color='red')
plt.title('Support Vectors Visualized')
plt.xlabel('X1')
plt.ylabel('X2')
plt.show()
```

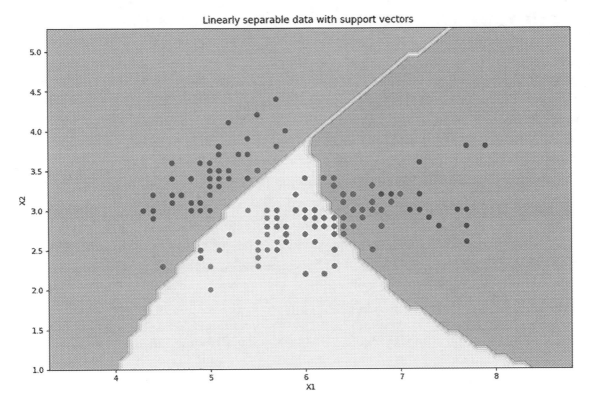

Figure 9-5. *Decision boundary in SVM*

Remember, we are currently limiting out capabilities to only two dimensions – the results will be much efficient if we take the complete data. However, it is easy to appreciate the robustness and flexibility of Support Vector Classifiers from this diagram.

Let's construct some more Support Vector Classifiers for the same dataset and try to compare them. In this example, we will take the first two classes (first 100 rows of Iris dataset) so that the boundaries will be clearer to visualize. This will allow us to compare the different implementations of SVCs.

```
X = df.iloc[:100,:2]
y = df.iloc[:100,5]
```

Here, X and y are coming from the inbuilt Iris dataset. For details, please see the previous example. Here, we have limited the dataset to first 100 rows and two columns.

Let's initialize the classifiers. We will create four objects, for linear, rbf, polynomial kernels, and one for LinearSVC implementation. We can adjust the required hyperparameters. An additional hyperparameter for you to tweak is C, which is the regularization parameter. It must be a positive number denoting the inverse of regularization strength that is applied as l2 penalty. We will currently leave it at its default value but keep it in the code for you to experiment with.

```
from sklearn import svm
C = 1.0  # SVM regularization parameter
svc = svm.SVC(kernel='linear', C=C).fit(X, y)
rbf_svc = svm.SVC(kernel='rbf', gamma=0.7, C=C).fit(X, y)
poly_svc = svm.SVC(kernel='poly', degree=3, C=C).fit(X, y)
lin_svc = svm.LinearSVC(C=C).fit(X, y)
```

In these lines, we have initialized and trained three SVC-based classifiers with kernels, namely, linear, rbf, and polynomial. We have another classifier using LinearSVC, which is a slightly different implementation.

```
titles = ['SVC with linear kernel',
          'LinearSVC (linear kernel)',
          'SVC with RBF kernel',
          'SVC with polynomial (degree 3) kernel']

xx,yy = np.meshgrid( np.arange(x_min, x_max, 0.1), np.arange(y_min,
y_max, 0.1)  )

plt.figure(figsize=(20,10))
for i, clf in enumerate((svc, lin_svc, rbf_svc, poly_svc)):
    plt.subplot(2, 2, i + 1)
    plt.subplots_adjust(wspace=0.4, hspace=0.4)

    Z = clf.predict(np.c_[xx.ravel(), yy.ravel()])

    # Put the result into a color plot
    Z = Z.reshape(xx.shape)
    plt.contourf(xx, yy, Z, cmap=plt.cm.coolwarm, alpha=0.4)

    # Plot also the training points
    plt.scatter(X.iloc[:, 0], X.iloc[:, 1], c=y, cmap=plt.cm.coolwarm)
    plt.xlabel('Sepal length')
    plt.ylabel('Sepal width')
    plt.xlim(xx.min(), xx.max())
```

```
plt.ylim(yy.min(), yy.max())
plt.xticks(())
plt.yticks(())
plt.title(titles[i])
```

```
plt.show()
```

This would print a chart showing the four SVC implementations (Figure 9-6).

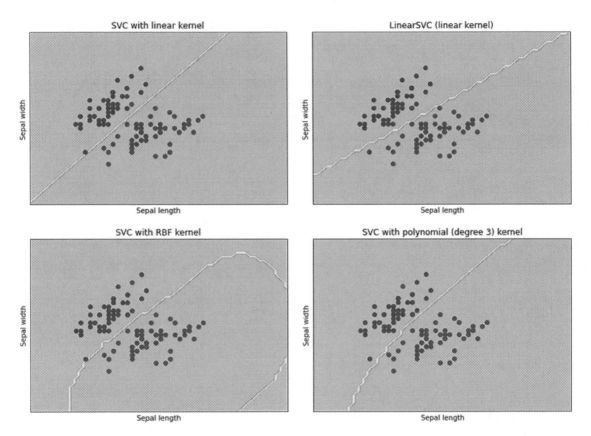

Figure 9-6. *Decision boundaries of four different Support Vector Classifiers on Iris dataset*

SVMs are highly effective in high dimension spaces. They are also very memory efficient due to the notion of support vectors, which is a relatively small subset of the training points. Kernel functions can help you make the decision boundary more flexible and robust. However, it is important to note that if the number of features is much greater than the number of samples, avoid overfitting in choosing kernel functions and regularization term is crucial.[2]

Summary

This chapter has presented advanced supervised learning techniques that have proven to be state of the art in various scenarios. In the next chapter, we will discuss another method that combines the power of multiple less accurate models to create an eventual robust model.

---

[2] https://scikit-learn.org/stable/modules/svm.html

CHAPTER 10

Ensemble Learning Methods

In the past three chapters, we have discussed and experimented with several supervised learning methods and learned how to evaluate them and tune their performance. Each class of algorithms has merits and demerits and suits a particular class of problems.

Ensemble learning, as shown in Figure 10-1, is a suite of techniques that uses multiple machine learning models in order to obtain better performance than we could have from any of the models. In many use cases, the individual models are called weak learners.

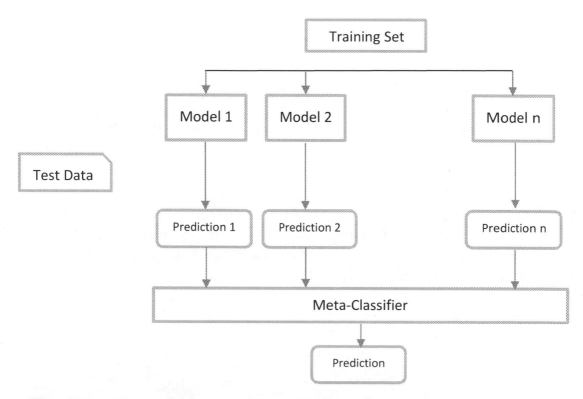

Figure 10-1. *Top-level schematic of ensemble learning methods*

A. Pajankar and A. Joshi, *Hands-on Machine Learning with Python*, https://doi.org/10.1007/978-1-4842-7921-2_10

In this chapter, we will learn techniques to combine the learning capabilities of small low-performance weak learners to create a high-performance and more accurate model.

Bagging and Random Forest

As we saw in the previous chapter, decision trees are constructed by selecting one of the attributes that provides the best separation of the classes. This is done either by computing the information gain or Gini index. Based on the splitting criteria, the dataset will follow one of the child nodes shown in Figure 10-2.

Color	Diameter	Label
Red	3	Apple
Yellow	3	Lemon
Purple	1	Grapes
Red	3	Apple
Yellow	3	Lemon
Purple	1	Grapes

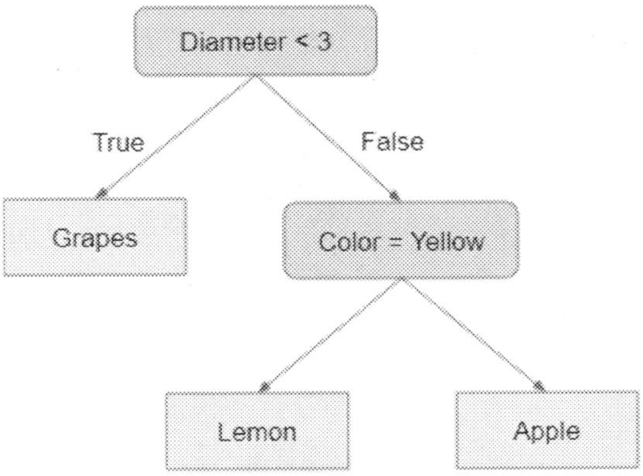

Figure 10-2. A simple decision tree based on a toy dataset

If we select a slightly smaller subset of the dataset, the tree thus constructed, even with the same splitting criteria and pruning rules, might be much different than the one compared earlier.

Figure 10-3 shows a new decision tree that was created by randomly selecting four (out of six) data points. The structure of the tree may totally change by changing the dataset as the computations of splitting criteria will change to reflect the statistics of the new training dataset. This is the idea behind one of the simple yet highly effective ensemble methods called random forest.

Color	Diameter	Label
Red	3	Apple
Yellow	3	Lemon
Purple	1	Grapes
Red	3	Apple
Yellow	3	Lemon
Purple	1	Grapes

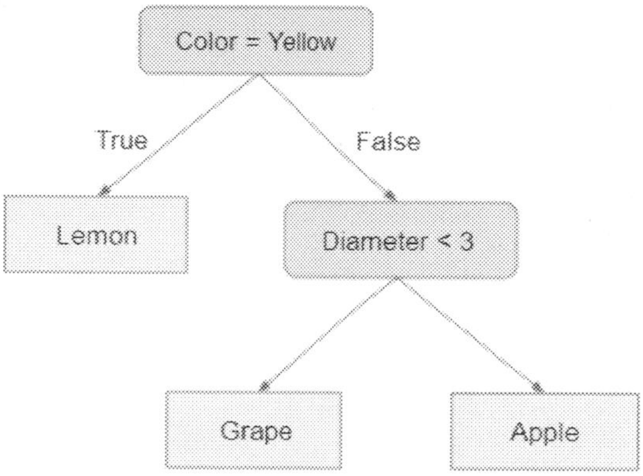

Figure 10-3. *Alternate decision tree created by randomly selecting a subset of training dataset*

Bagging, or Bootstrap AGGregating, creates an ensemble model through bootstrapping and aggregation processes. Instead of creating a model on the whole training dataset, we pull a sample of the training set and use it for training instead. The sample is generated randomly with each training data point having the equal probability of being sampled. If N represents the size of training dataset and M is the number of random data points to be considered for training, and M<N, each point is randomly sampled with replacement; that is, it can occur more than once in the subset. This also means there is a probability that a data point might not occur in the training dataset at all.

In the training process shown in Figure 10-4, we create k models, where k is a predefined number and a hyperparameter that is easy to configure. Each tree is constructed independently, and thus, the process can be implemented in parallel. Eventually, we will have k-independent decision trees with possibly different structures, and they might give different results for the same test item.

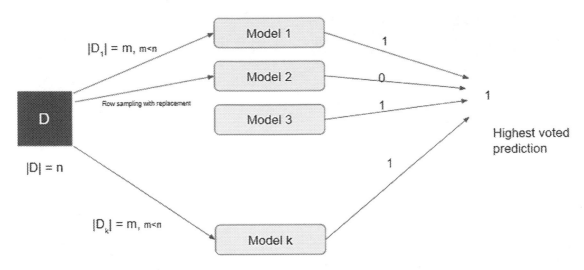

Figure 10-4. *Bagging or bootstrap aggregation process*

During the prediction phase, the test item runs through each tree and is assigned a label by the decision tree. The labels are then aggregated and voted – and the label with the highest number of votes across the k trees is considered as the final output. In this example, if k=4 and the class labels predicted by the four models are 1, 0, 1, and 1, there are three votes for 1 and one vote for 0. Thus, the final class that is assigned is 1.

Random Forest in Python

In this short example, we aim to construct a random forest for classification of Iris dataset and visualize the individual trees that are a part of the ensemble. Let's begin by preparing the dataset for our experiment.

```
import pandas as pd
from sklearn.datasets import load_iris
iris = load_iris()
df = pd.DataFrame(iris.data, columns=iris.feature_names)

from sklearn.model_selection import train_test_split
X_train, X_test, y_train, y_test = train_test_split(df, iris.target,
test_size=0.3)
```

In Scikit-learn, sklearn.ensemble package contains implementations of various ensemble methods for classification, regression, and anomaly detection. One of the options is the Bagging meta-estimator, which accepts another weak classifier as an argument and builds several instances of the weak classifier on random samples of the training set. This treats the internal weak-classifier algorithm as a black box and doesn't require any changes in how the internal models work.

To create a random forest, we will use RandomForestClassifier, which allows decision tree–specific hyperparameters that can be easily tuned externally. It learns a large number of decision tree classifiers on various random samples of the dataset in order to improve the quality of predictions.

We can specify Bagging specific hyperparameters, like max_samples, which is used to specify the number or ratio of training data points that will be randomly selected in each iteration. Besides, we can specify n_estimators, which is the number of individual trees that are created. Apart from these, other hyperparameters like max_depth, min_samples_split, criteria, etc., can be used the way they are used while constructing a decision tree.

Let's initialize a random forest with ten trees and train it.

```
from sklearn.ensemble import RandomForestClassifier
clf = RandomForestClassifier(n_estimators=10, max_samples=0.7)
clf.fit(X_train, y_train)
```

The constituent decision trees can be visualized through clf.estimators_. We can iterate over them and visualize the structure of the trees using Graphviz and PyDotPlus.

```
from IPython.display import Image
from sklearn.tree import export_graphviz
import pydotplus

for i, estimator in enumerate(clf.estimators_):
    dot_data = export_graphviz(estimator)
    graph = pydotplus.graph_from_dot_data(dot_data)
    graph.write_png('tree'+str(i)+'.png')
    display(Image(graph.create_png()))
```

Some of the trees are shown in Figure 10-5. In this code block, we also added a line that uses write_png() to save each tree in the hard drive.

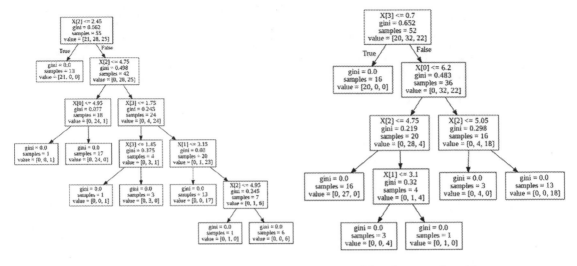

Figure 10-5. *Example of some of the trees generated by the random forest*

As each decision tree in the Bagging process is independent of each other, Scikit-learn implementation of random forest also provides an option to construct the trees in parallel. This is possible by passing the n_jobs parameter. By default, it is set to None, thus no parallelization.

Boosting

Boosting is another highly popular ensemble learning technique. Instead of constructing independent trees through random samples of the training data, boosting process constructs the trees in an iterative process.

In boosting, the ensemble is created incrementally by training a new model from a subset of the training data that considers a more proportion of data points that the previous models misclassified. In this section, we will discuss a boosting method called AdaBoost that incrementally builds decision trees using this method.

AdaBoost further simplifies the decision trees by constructing decision stumps, which is a weak learner by itself, but the boosting process ultimately combines such weak learners to create a more accurate classifier. A decision stump, as shown in Figure 10-6, is a simple decision tree consisting of only one split, thus a height of one. This means each classifier splits the examples into two subsets using only one feature.

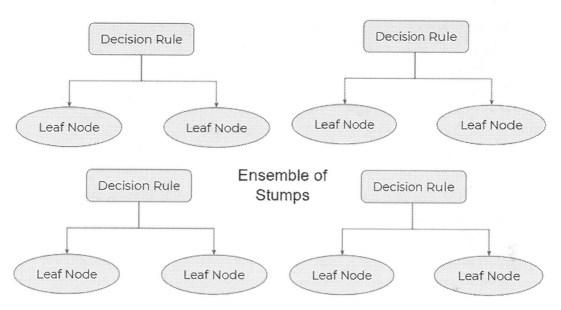

Figure 10-6. *Decision stumps, the simplest form of decision trees*

Each stump is a very simple classifier with a linear and straight decision boundary. Thus, a stump can't do much by itself. However, as shown in Figure 10-7, the process of sampling and assigning weights to each stump individually is what improves the end results.

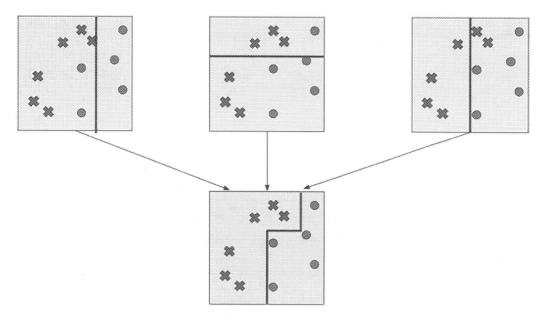

Figure 10-7. *Decision boundaries created by each decision stump combined to obtain a more complex decision boundary*

Say, we initially assign a sampling weight of 1/N to all the rows of the training data. They are sampled based on the weight and used to construct a decision tree or a decision stump called Model 1.

Suppose we want to boost the accuracy of a learning method. We are given D, a dataset of d class-labelled tuples, $(X_1, y_1), (X_2, y_2), \ldots, (X_d, y_d)$, where y_i is the class label of tuple X_i. Initially, AdaBoost assigns each training tuple an equal weight of 1/d. Generating k classifiers for the ensemble requires k rounds through the rest of the algorithm.

In round i, the tuples from D are sampled to form a training set, D_i, of size d. Sampling with replacement is used – the same tuple may be selected more than once. Each tuple's chance of being selected is based on its weight. A classifier model, M_i, is derived from the training tuples of D_i. In Figure 10-8, Model 1 is generated from the first set of samples where each sample from the training dataset carried equal weight.

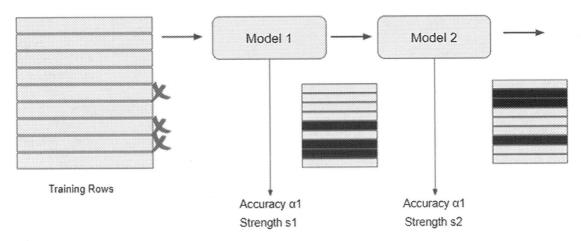

Figure 10-8. *Process of creating an ensemble model using AdaBoost*

Now that we have a simple classification model ready, we test the model using the same dataset D_i we used for training. The weights of the training tuples are then adjusted according to how they were classified. If a tuple was incorrectly classified, its weight is increased. If a tuple was correctly classified, its weight is decreased. A tuple's weight reflects how difficult it is to classify – the higher the weight, the more often it has been misclassified. These weights will be used to generate the training samples for the classifier of the next round. An overview of this process is shown in Figure 10-8.

Each model's accuracy is directly used as the notion of strength of the classifier which will be used while predicting.

While predicting classification results on unseen data, instead of voting the results of all the models like in Bagging methods, we calculate the weighted average of the individual models.

We initialize the weight of each possible class to 0. We get the weight of each individual model by considering the error of the model using

$$w_i = log \frac{1 - error(M_i)}{error(M_i)}$$

If i_{th} model gives prediction as class c and has weight of w_i, we add w_i to the weight for class c in overall prediction. After predicting the label from all the classes, we assign the class label with the highest weight as the final prediction.

Sometimes, the resulting "boosted" model may be less accurate than a single model derived from the same data. Bagging is less susceptible to model overfitting. While both can significantly improve accuracy in comparison to a single model, boosting tends to achieve greater accuracy.

AdaBoost is easy to implement. It iteratively corrects the mistakes of the weak classifier and improves accuracy by combining weak learners. You can use many base classifiers with AdaBoost. AdaBoost is not prone to overfitting. This can be found out via experiment results, but there is no concrete reason available. However, it is highly affected by outliers because it tries to fit each point perfectly. It is particularly vulnerable to uniform noise.

Boosting in Python

For clearer comparison, we will run AdaBoost for the same training dataset we used in the previous example. We have already extracted the dataset and split into `X_train`, `X_test`, `y_train`, and `y_test`.

In Scikit-learn, `sklearn.ensemble.AdaBoostClassifier` accepts a base estimator, which defaults to a one-level decision tree (stump). However, we can initialize a decision tree with different hyperparameters, which is then passed to the AdaBoostClassifier. We can configure the number of weak learners using n_estimators and a learning rate using learning_rate, which configures the weight applied to each classifier during each iteration. A higher learning rate increases the contribution of each individual classifier. Its default value is 1.

```
from sklearn.ensemble import AdaBoostClassifier
from sklearn.tree import DecisionTreeClassifier

base = DecisionTreeClassifier(criterion='gini', max_depth=1)
model_ada = AdaBoostClassifier(base_estimator = base, n_estimators=10)
model_ada.fit(X_train, y_train)
```

Remember, specifying a base_estimator is optional. Once the model fits, we can see the constituent estimators using model_ada.estimators_.

If you wish to visualize the structure of individual stumps, you can use the method using PyDotPlus in the previous section. These will be simple decision trees containing only one split each.

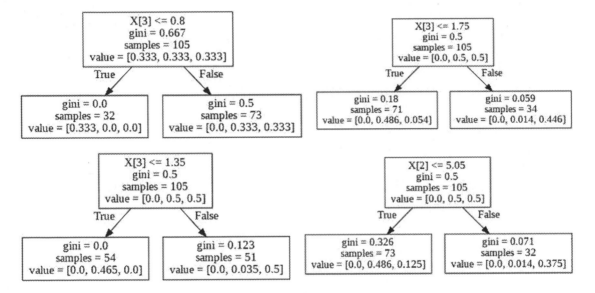

Figure 10-9. *Decision stumps created by the AdaBoost model*

Based on the constituent estimators as shown in Figure 10-9, AdaBoost object can give the overall importance of each feature in the training dataset. In Iris dataset, we have four features. The following code will output a list of four values, indicating the importance of the four columns.

```
model_ada.feature_importances_
array([0.1, 0. , 0.3, 0.6])
```

The idea behind AdaBoost is to use very simple weak learners and reduce the error rate to achieve a much accurate model. We can assign a base estimator instead of using the default stump. Let's run an experiment on synthetic dataset and visualize the impact of base_model on the overall accuracy.

We will construct a synthetic dataset using the sklearn.dataset.make_classification method. Here's a simple example.

```
import matplotlib.pyplot as plt
from sklearn.datasets import make_classification

plt.figure(figsize=(8, 8))
X, y = make_classification(n_samples = 1000, n_features=2, n_redundant=0,
n_informative=2, n_clusters_per_class=1, n_classes=3)
plt.scatter(X[:, 0], X[:, 1], marker='o', c=y, s=25, edgecolor='k')
```

Here, we instruct Scikit-learn to construct a dataset containing 1000 samples with two features each and spread the data into three clusters, each denoting a class. A simple visualization shows that it is much complex compared to the Iris dataset – however, there is a fuzzy separation between the three classes. The output of this code block is shown in Figure 10-10.

Due to random generation, your data might look different.

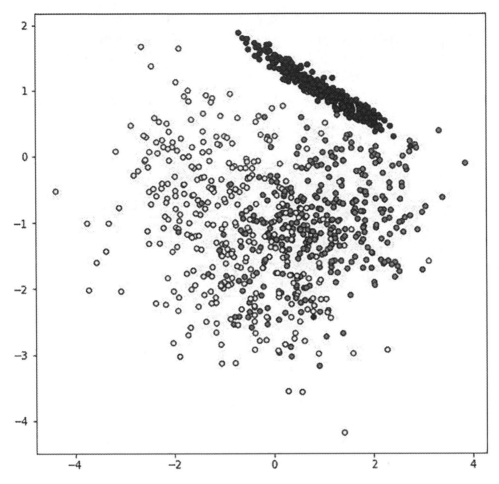

Figure 10-10. *Scatter plot of randomly generated dataset*

We will use this method to generate a more complex dataset and train an AdaBoost model out of these.

```
from sklearn.model_selection import train_test_split
X, y = make_classification(n_samples = 3000, n_features=10, n_redundant=0,
n_informative=10, n_clusters_per_class=1, n_classes=3)
X_train, X_test, y_train, y_test = train_test_split(X, y, test_size=0.3)
```

In these lines, we first create a random classification dataset with 3000 samples, each data point in ten columns spread over three classes present in their own clusters. We split the dataset into training and test set so that the reported accuracy measures come independently from the test data.

Let's import the requirements. In this example, we will explicitly create objects of DecisionTreeClassifier to set the depth at different levels and analyze its impact on the overall accuracy.

A list called `accuracy_history` will store the accuracy of multiple models for visualization and comparison.

```
from sklearn.ensemble import AdaBoostClassifier
from sklearn.tree import DecisionTreeClassifier
from sklearn.metrics import accuracy_score

accuracy_history = []
```

Now we want to create multiple AdaBoost classifiers, each containing the same number of estimators, that is, 50. However, in each iteration, the depth of the constituent decision tree will be different.

```
for i in range(1, 20):
    tree = DecisionTreeClassifier(max_depth = i)
    model_ada = AdaBoostClassifier(base_estimator=tree, n_estimators=50)
    model_ada.fit(X_train, y_train)
    y_pred = model_ada.predict(X_test)
    accuracy_history.append(accuracy_score(y_test, y_pred))
```

Thus, in the first iteration, we create an AdaBoostClassifier consisting of 50 stumps, or decision trees with a depth of 1. In the next iteration, we create another AdaBoostClassifier consisting of 50 decision trees with a depth of 2. The 20th AdaBoostClassifier will consist of 50 decision trees with a depth of 19.

Will the classifier with better and more complex trees have significant improvement over the classifier with simpler trees? To answer that, we will find the predictions on test dataset in each iteration and store the accuracy history in the list. Let's visualize the accuracy history.

```
plt.figure(figsize=(8, 4))
plt.plot(accuracy_history, marker='o', linestyle='solid', linewidth=2,
markersize=5)
plt.grid(True)
plt.xticks(range(1,20))
plt.ylim((0.5,1))
```

```
plt.xlabel("Depth of weak-learner")
plt.ylabel("Accuracy Score")

plt.show()
```

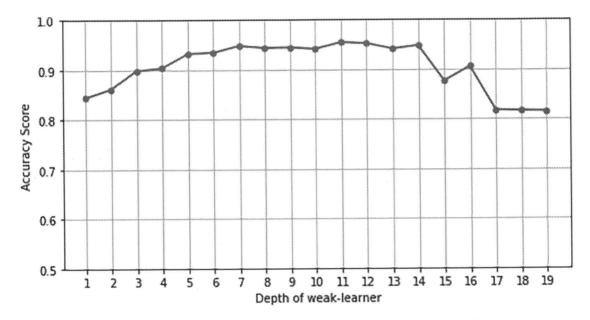

Figure 10-11. *Accuracy of model based on the depth of constituent learners*

We observe in Figure 10-11 that there's an increase in accuracy in the beginning as we make the constituent trees more complex – however, beyond a certain depth, there is no improvement in accuracy. In some practical scenarios, stumps or very simple decision trees can provide good enough accuracy for relatively low computation requirements.

Stacking Ensemble

Stacking, or stacked generalization, is another ensemble learning technique that combines the predictions from multiple machine learning models by assigning the weights on individual constituent classifiers.

This is significantly different from Bagging – which is the approach of creating the same type of model (e.g., decision trees) on different samples of the training dataset. Here, we may choose different kinds of estimators with different hyperparameters. In simple words, a stacking ensemble builds another learning layer (meta-model) that learns how to combine the predictions of constituent models.

Implementations of stacking ensembles differ based on the type of machine learning being solved. If we have multiple regression models trained on the same dataset, the meta-model will take the outputs of the individual regression models and try to learn the weight of each model to produce the final numeric output.

Stacking in Python

Scikit-learn provides an easy to use implementation for stacking estimators using `sklearn.ensemble.StackingClassifier`. It accepts a list of base models (estimators) that will be stacked together. There's a final estimator (default logistic regression) that tries to learn to predict the final value based on the output of constituent estimators. The final estimator is trained using cross-validated predictions of the base estimators.

```
from sklearn.model_selection import train_test_split

X, y = make_classification(n_samples = 3000, n_features=10, n_redundant=0,
n_informative=10, n_clusters_per_class=1, n_classes=3)
X_train, X_test, y_train, y_test = train_test_split(X, y, test_size=0.3)
```

Let's import classifier models for logistic regression, KNN, decision trees, support vector, and naïve Bayes.

```
from sklearn.linear_model import LogisticRegression
from sklearn.neighbors import KNeighborsClassifier
from sklearn.tree import DecisionTreeClassifier
from sklearn.svm import SVC
from sklearn.naive_bayes import GaussianNB
```

To keep this example simple, let's create a list of tuples – with each tuple containing the name and the estimator object for each classifier. We will consider the default values of all hyperparameters.

```
models = [('Logistic Regression',LogisticRegression()),
          ('Nearest Neighbors',KNeighborsClassifier()),
          ('Decision Tree',DecisionTreeClassifier()),
          ('Support Vector Classifier',SVC()),
          ('Naive Bayes',GaussianNB())]
```

Let's train and evaluate them in a loop and plot the accuracy of the five models. The output of this block is shown in Figure 10-12.

```
accuracy_list = []

for model in models:
    model[1].fit(X_train, y_train)
    y_pred = model[1].predict(X_test)
    accuracy_list.append(accuracy_score(y_test, y_pred))

plt.figure(figsize=(8, 4))
model_names = [x[0] for x in models]
y_pos = range(len(models))
plt.bar(y_pos, accuracy_list, align='center', alpha=0.5)
plt.xticks(y_pos, [x[0] for x in models], rotation=45)
plt.ylabel('Accuracy')
plt.title('Comparision of Accuracies of Models')
plt.show()
```

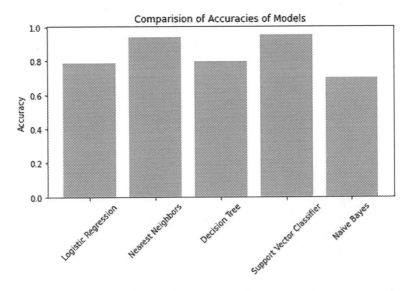

Figure 10-12. *Comparison of accuracies of different models*

Instead of being able to use one of the well-performing models, we can create a stacking classifier to act as an ensemble of all the models. We will specify the five models we initialized earlier and a logistic regression as the final estimator, which will use the results of five estimators as input and learn the weights that must be used to assign the final class.

```
from sklearn.ensemble import StackingClassifier
stacking_model = StackingClassifier(estimators=models,
final_estimator=LogisticRegression(), cv=5)
stacking_model.fit(X_train, y_train)
```

This will initiate training for each of the constituent estimators. To predict, we simply need to predict using stacking_model.

```
y_pred = stacking_model.predict(X_test)
accuracy_score(y_test, y_pred)
```

The accuracy score we thus obtain will usually be higher than the highest accuracy we obtained on any individual model. We would like to print another bar chart comparing the accuracies of all the models.

```
accuracy_list.append(accuracy_score(y_test, y_pred))
model_names = [x[0] for x in models]
model_names.append("Stacked Model")
plt.figure(figsize=(8, 4))
y_pos = range(len(model_names))
plt.bar(y_pos, accuracy_list, align='center', alpha=0.5)
plt.xticks(y_pos, model_names, rotation=45)
plt.ylabel('Accuracy')
plt.title('Comparision of Accuracies of Models')
plt.show()
```

Figure 10-13 shows that the stacked model definitely performs better than any of the individual models.

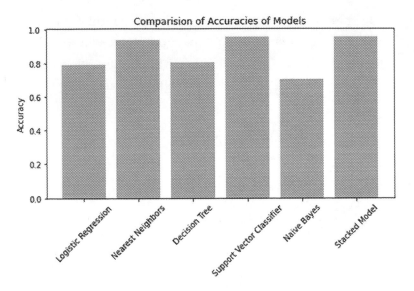

Figure 10-13. *Comparison of accuracies of individual models with respect to the stacked model*

Summary

In this chapter, we explored several techniques to combine relatively weak learners to create a final model that is much more accurate than the individual models. Bagging, boosting, and stacking are three approaches that are highly used in combining various estimators.

In the next chapter, we will study a different class of machine learning problems called unsupervised learning.

CHAPTER 11

Unsupervised Learning Methods

So far, we've discussed several solutions for either predicting a continuous variable based on given independent variables or predicting what class or classes a particular data item belongs to. We also discussed a few methods to combine multiple models to create a more effective meta-model. All these methods require a training dataset consisting of the values or labelled that are expected as the predicted output, thus, these leading to the name of supervised learning. In this section, we will discuss a different class of machine learning problems and solutions in which rather than predicting, the aim is to either transform the data or discover patterns without the need of a set of explicit training labels.

We will discuss three major types of unsupervised learning methods:

1. Dimensionality reduction

2. Clustering

3. Frequent pattern mining

Dimensionality Reduction

Dimensionality reduction refers to a set of techniques used to summarize the data in a reduced number of dimensions. One common application of converting the data to a reduced set of dimensions is visualization. In the previous chapters, we saw that Iris dataset has four independent variables – sepal width, sepal length, petal width, and petal length. If we wish to plot the 150 data points to understand the distribution, we picked two of these columns and ignored the other two. However, such an approach hides the patterns in which, say, two different flowers have similar sepal width but

185

© Ashwin Pajankar and Aditya Joshi 2022
A. Pajankar and A. Joshi, *Hands-on Machine Learning with Python*, https://doi.org/10.1007/978-1-4842-7921-2_11

drastically different petal width that differentiates between the two. Dimensionality reduction methods provide us an alternate way to transform the data into a new set of two dimensions such that the differences or the variances in the original dataset are maximized.

Another common use of dimensionality reduction is to preprocess the data before further machine learning experiments to simplify the structure and avoid the curse of dimensionality. Such methods simplify the dataset while still preserving the intrinsic structures and patterns.

Understanding the Curse of Dimensionality

You'll often encounter data that is present in high-dimensional spaces. In simple forms, there might be too many columns in the data unlike the examples we've seen so far. There might be effects of one-hot-encoding or textual feature extraction like n-grams, which can lead to a large number of resulting dimensions. When the numbers of dimensions increase, the data becomes too sparse – that is, the number of columns increases with most of them consisting of insignificant values (mostly zeros). Meanwhile, the number of data samples, or the number of rows, stays the same. Let's take the example in Figure 11-1 and identify the patterns.

Say, you are working with a dataset (left) that contains details of employees consisting of the cities they reside in and their monthly salary (converted to US dollars).

Emp No	City	Salary	Emp No	Banglore	Hyderabad	Delhi	San Francisco	Santa Clara	Salary
1101	Bangalore	900	1101	1	0	0	0	0	900
1102	San Francisco	6500	1102	0	0	0	1	0	6500
1103	Hyderabad	1250	1103	0	1	0	0	0	1250
1104	Santa Clara	8000	1104	0	0	0	0	1	8000
1105	Delhi	1150	1105	0	0	1	0	0	1150
1106	Bangalore	1200	1106	1	0	0	0	0	1200

Figure 11-1. *Sample dataset with city of the employees present in a one-hot representation*

The city column is encoded using one-hot encoding as shown in Figure 11-1, thus expanding to five columns, each representing a unique city in the database. Although the pattern is easy to understand if you know which country each city is from, but without additional information, it is hard to find such patterns in so sparse data.

Principal component analysis (PCA) is one of the most simple and most common techniques applied for dimensionality reduction. It transforms the data into a lesser number of dimensions that are almost as informative as the original dataset.

Principal Component Analysis

The process of performing principal component analysis begins by first computing the covariances of all the columns and storing in a matrix, which summarizes how all the variables are related to each other. This can be used to find eigenvectors, which show the directions in which the data is dispersed, and eigenvalues, which show the magnitude of the importance along each eigenvector.

Figure 11-2 shows a set of data points originally plotted in two dimensions: X and Y. PCA tries to project these points on a dimension that leads to maximum variance, or maximum spread of the data. If we project these points on X axis, we see that all the points are spread between x1 and x2. The variance will capture the spread of the projection of the points on x axis. Similarly, if we project the points on Y axis, they'll be spread between y1 and y2. If we are able to project in an entirely different axis, denoted as K axis in Figure 11-2, we observe that the spread between k1 and k2, as well as the variance observed, is higher compared to the one we obtained from projections on X and Y axes. If PCA tries to find one dimension on which to project the dataset, it will pick this dimension. This is the first principal component. If we wish to choose the second principal component, we have to choose the axis orthogonal to this. The aim here is to first find the principal components and, second, provide a transformation for mapping original data to the principal components.

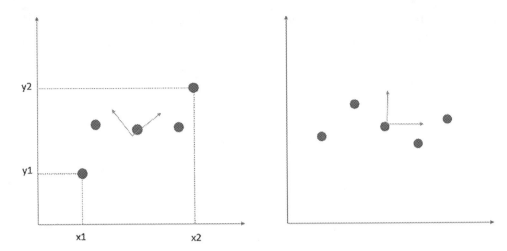

Figure 11-2. *Transformation of original dataset along the principal components*

To standardize the data to perform PCA, we first compute a d-dimensional mean vector; that is, compute the mean for each column. This is used to standardize the data and bring all the columns on a similar scale. This helps avoid the cases where a column spread over a large range, say, salary, will dominate over a column spread in a smaller range, like GPA. Standardization will bring all the columns to a similar scale.

$$z = \frac{value - mean}{standard\ deviation}$$

This is followed by computing a covariance matrix that provides a way to analyze the relationship between the columns. A covariance matrix gives the covariances across each combination of the dimensions in the dataset. Say, for a dataset with three columns, d=3, referred to as x, y, and z, the covariance matrix is given by

$$\begin{bmatrix} cov(x,x) & cov(x,y) & cov(x,z) \\ cov(y,x) & cov(y,y) & cov(y,z) \\ cov(z,x) & cov(z,y) & cov(z,z) \end{bmatrix}$$

where Cov(x,y) represents the covariance of x and y. If this is positive, it means that x and y are positively correlated; that is, their values increase or decrease together. If it is negative, it means that they are negatively correlated. If the value of x increases, the value of y decreases.

We then compute the eigenvectors and eigenvalues of the covariance matrix to identify the principal components. The aim of PCA is to find the principal components that explain the maximal amount of variance. The eigenvector corresponding to the highest eigenvalue is the first principal vector. We can decide how many principal components we want and select the two eigenvectors accordingly. If we wish to use PCA for two-dimensional visualization, we will pick the top two.

To transform the dataset along the principal component analysis, we use the matrix of eigenvectors and multiply it with the standardized original dataset.

Principal Component Analysis in Python

Due to simplicity, PCA can be performed using basic linear algebra functions by following the steps mentioned in the previous section. Scikit-learn also offers an even easier solution to apply PCA using the style of API consistent with other operations in sklearn. Let's see how we can use it to perform PCA in the Iris dataset and use it to visualize it from an entirely different perspective. We will resolve the four features of IRIS dataset into two principal components, which can be directly mapped on a 2D chart. Let's import the dataset using the methods we used in the previous chapter.

```
from sklearn import datasets
iris = datasets.load_iris()
X = iris.data
y = iris.target
```

As you know, there are four columns in the dataset – and it is hard to visualize on a screen. The best we can do at this level is visualize three dimensions (and ignore the fourth).

```
from mpl_toolkits.mplot3d import Axes3D
import matplotlib.pyplot as plt
fig = plt.figure(1, figsize=(4, 3))
ax = Axes3D(fig, rect=[0, 0, .95, 1], elev=48, azim=134)
ax.scatter(X[:, 0], X[:, 1], X[:, 2], c=y, cmap=plt.cm.nipy_spectral,
edgecolor='k')
```

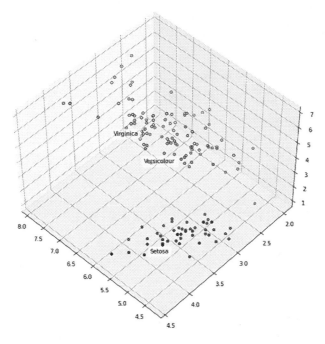

Figure 11-3. *Iris dataset represented in 3D axes*

You can print class labels using ax.text3D().

```
for name, label in [('Setosa', 0), ('Versicolour', 1), ('Virginica', 2)]:
    ax.text3D(X[y == label, 0].mean(),
              X[y == label, 1].mean() + 0.5,
              X[y == label, 2].mean(), name,
              horizontalalignment='center',
              bbox=dict(alpha=.5, edgecolor='w', facecolor='w'))
```

```
plt.show()
```

The plot in three dimensions as shown in Figure 11-3 conveys a little more information than the charts we've seen before. However, there is a possibility that there's something more the missing dimension must convey. PCA will help us translate the data to two dimensions while making sure that the pattern of distributions is preserved.

To perform PCA, we need to import the required classes and apply fit and transform. This will convert the dataset to the new two-dimensional space.

```
from sklearn import decomposition
pca = decomposition.PCA(n_components=2)
pca.fit(X)
X = pca.transform(X)
```

You can check the first few rows of X after transforming

```
X[:5]
```

The output shows that X is now a two-column matrix with shape (150,2). The interesting observation here is the values of X don't actually correspond to the actual sepal and petal lengths or widths. These are simply transformations in the new two-dimensional space.

```
>> array([[-2.68412563,  0.31939725],
       [-2.71414169, -0.17700123],
       [-2.88899057, -0.14494943],
       [-2.74534286, -0.31829898],
       [-2.72871654,  0.32675451],
```

We can visualize using Matplotlib with data points from each representative class (variety of Iris flower) plotted in a different color.

```
fig = plt.figure(figsize=(8,8))
```

```
plt.scatter(X[:,0], X[:,1], c=y, cmap=plt.cm.nipy_spectral, edgecolor='k')
```

```
for name, label in [('Setosa', 0), ('Versicolour', 1), ('Virginica', 2)]:
    plt.text(X[y == label, 0].mean(),  X[y == label, 1].mean(),
    name,  horizontalalignment='center', bbox=dict(alpha=0.8,
    edgecolor='w', facecolor='w'))
```

```
plt.show()
```

This will show the scatter plot of the PCA transformed version of Iris dataset in two dimensions. The plot in Figure 11-4 shows more realistic interaction between the two classes – Versicolor and Virginica. In this plot, you can see that a simple transformation like this across the axes of highest variance leads to a better simplification of the dataset.

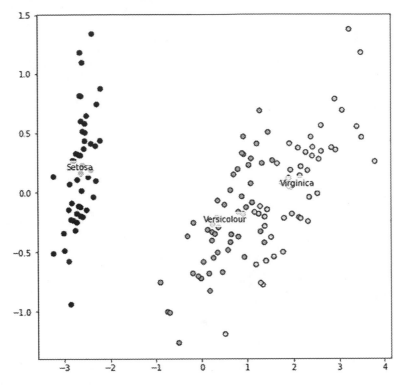

Figure 11-4. *Iris dataset represented along two principal components*

Massive datasets are increasingly widespread in all sorts of disciplines. To interpret such datasets, we need to decrease dimensionality to preserve the most highly related data. We can use PCA to reduce the number of variables, avoid multicollinearity, or have too many predictors relative to the number of observations.

Feature reduction is an essential preprocessing step in machine learning. Therefore, PCA is an essential step of preprocessing and very useful for compression and noise removal in the data. It reduces the dimensionality of a dataset by finding a new set of variables smaller than the original set of variables.

Clustering

Clustering is a suite of simple yet highly effective unsupervised learning methods that help group the data into meaningful groups that reveal an underlying pattern. Imagine you are asked by the Director of Human Resources of your organization to divide all the employees into five groups – so that the company's leadership can determine which training programs should the employees in each group be signed up for. In this kind of problem, you want to create the groups so that (1) all the employees in one group should be more or less similar to one another and (2) employees in different group should be significantly different from each other. Such problems, where you need to group the data into distinct segments, are called clustering.

In this problem, we do not have predefined categories of employees. We do not have a label based on which the algorithm learns to find the mapping, as we saw in several classification algorithms. Here, the algorithm learns to partition the data based on distance measures.

In the following subsections, we will first learn a very popular clustering method and look at its implementation in Python. We will also discuss another application of clustering as one of the steps in AI applications that leverage image processing or computer vision. We will then study another clustering method that can detect clusters of arbitrary shape.

Clustering Using K-Means

K-means is a centroid-based cluster partitioning method. K-means aims to create k clusters and assign data points to them so that there is high intracluster similarity and low intercluster similarity. The algorithm is based on the concept of minimizing the inertia, or within-cluster sum of squares. Inertia is given by

$$\sum_{i=0}^{n} \min_{\mu_j \in C} \left(\| x_i - \mu_j \|^2 \right)$$

The algorithm begins by selecting K random points, which will be hereby referred to as centroids of the resulting clusters. The algorithm then proceeds to multiple iterations of refining the cluster.

In each iteration, for each data point, the algorithm finds the nearest centroid by computing the distances of the point with all the centroids and assigns it to that cluster. Once all the points are assigned to the clusters, a new centroid is calculated by computing the mean of all the points (across all the dimensions). This process continues until there is no change of cluster centers or cluster assignments, or after a fixed number of iterations are over.

The algorithm begins with randomly initialized cluster centers. Due to this, a good convergence might not always happen. In practice, it is a good idea to repeat the clustering algorithm multiple times with randomly initialized cluster centers and sample the final cluster distributions.

K-Means in Python

Scikit-learn provides an efficient implementation of K-means.

By default, the initial selection of the cluster centers is based on an algorithm called **k-means++**. This algorithm chooses one cluster center uniformly at random among the data points. Iteratively, the distance between all the points and the center is calculated. The next cluster center is chosen using a weighted probability distribution where a point is chosen with probability proportional to the distribution of squared distances. This is repeated till k points are chosen. This method often leads to significant improvement in the final performance of k-means. It is also possible to supply the k center points directly, if you decide to.

Let's generate a synthetic dataset that contains some kind of cluster shapes. Scikit-learn contains such functions in `sklearn.datasets`. Figure 11-5 shows a scatter plot showing clearly interpretable clusters of data.

```python
from sklearn.datasets import make_blobs
X, y = make_blobs(n_samples=500, centers=5, n_features=2, random_state=2)
plt.scatter(X[:,0], X[:,1], edgecolor='k')
```

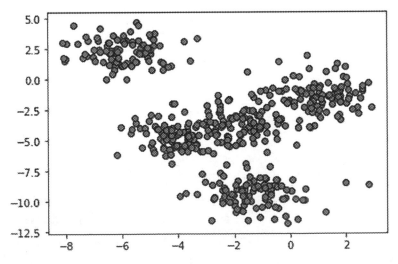

Figure 11-5. *Randomly generated data present in clusters*

To perform K-means clustering, we need to import KMeans from sklearn.clustering. The usage is similar to the standard sklearn functions for supervised learning.

```
from sklearn.cluster import KMeans
kmeans = KMeans(n_clusters=5)
kmeans.fit(X)
```

Here, the last line instructs the algorithm to learn the clustering based on the data that is provided. To find the cluster labels for each point, we can use the predict method.

```
y = kmeans.predict(X)
```

Alternatively, you can fit and predict in one line using

```
y = kmeans.fit_predict(X)
```

y will be a one-dimensional array containing a number from 0 to 4, each representing a cluster out of the five that were thus generated. You can obtain the cluster centers using

```
kmeans.cluster_centers_
>> array([[-3.99782236, -4.6960255 ],
       [-5.92952036,  2.24987809],
       [ 1.08160065, -1.26589927],
       [-1.28478773, -9.3769836 ],
       [-1.48976417, -3.56531061]])
```

Because the data is in two dimensions, there are five two-dimensional cluster centers. The points are simply assigned based on the cluster center closest to them.

Let's see the clusters that are generated. We will create a scatter plot for the 500 points we generated, and we will assign them a color according to the cluster they are in. We will also plot the cluster centers with a different color (black) and marker.

```
plt.scatter(X[:,0], X[:,1], c=y)
plt.scatter(kmeans.cluster_centers_[:,0], kmeans.cluster_centers_[:,1],
c='black', marker='+')
```

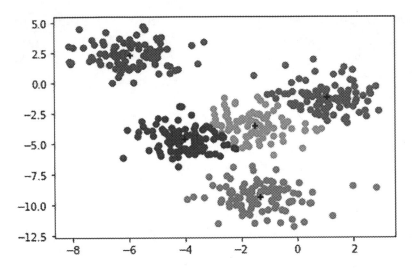

Figure 11-6. *Five clusters produced by K-means*

The middle blob is divided into three parts based on this clustering model as shown in Figure 11-6. However, intuitively, you can see that these parts are not very clearly separable, and thus, there should be only one larger cluster that combines these three parts. Let's see what kind of clusters are formed if we generate only three of them.

```
kmeans = KMeans(n_clusters=3)
y = kmeans.fit_predict(X)

plt.scatter(X[:,0], X[:,1], c=y)
plt.scatter(kmeans.cluster_centers_[:,0], kmeans.cluster_centers_[:,1],
c='black', marker='+')
```

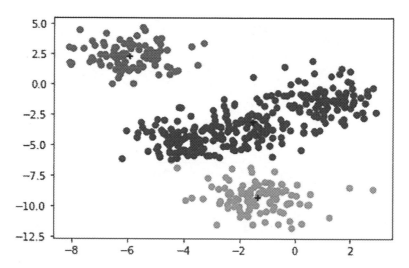

Figure 11-7. *Three clusters produced by K-means*

As we expected, Figure 11-7 shows that the larger blob in the middle becomes one cluster as a whole, while the other two clusters stay almost the same. However, sometimes, we might need better ways to define which of them is better. We'll discuss this in the next section.

What Is the Right K?

K-means algorithm expects a predetermined value of k. In some cases, like the one we discussed in the introduction, we might already have a predetermined value of k due to the business reasons or domain knowledge. However, if there is no deliberate reason to have a predetermined k, it is better to create using K-means for different values of k and analyze the quality of the clusters using the notion of purity or the error. As we increase the number of k, the error will always decrease – however, we will notice a value of k (or a couple of values) where there's a significant decrease. This value is called a knee point and decided as a safe value.

```
error = []
for i in range(1,21):
    kmeans = KMeans(n_clusters=i).fit(X)
    error.append(kmeans.inertia_)

import matplotlib.pyplot as plt
```

```
plt.plot(range(1,21), error)
plt.title("Elbow Graph")
plt.xlabel("No of clusters (k)")
plt.ylabel("Error (Inertia)")
plt.xticks([1,2,3,4,5,6,7,8,9,10])
plt.show()
```

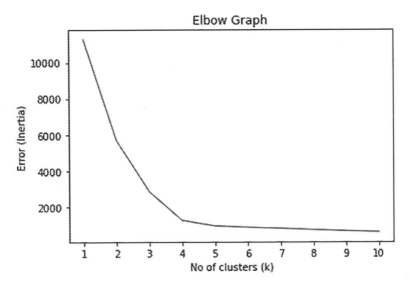

Figure 11-8. *Error rate reduces as we increase the number of clusters*

The graph in Figure 11-8 shows that the error in terms of inertia reduces as we increase the number of clusters. Inertia represents how internally coherent the clusters are. K-means algorithms is based on minimizing the inertia. This reduction of error (or improvement of cluster quality) is prominent when the number of clusters increases from 1 to 2, 3, 4, and 5. Beyond that point, the improvement is minimal. This point, called the knee point, is a good indicator of how many clusters should be found. Although it is not clearly obvious where the knee point lies, we usually select it by eye using the elbow plot and analyze the summary statistics that evaluate the quality of the clusters produced by the selected k (say, 3). We might repeat this using a different k (4) and see which k leads to mode meaningful results. From the comparison, it looks like for our dataset, 3 is an ideal number of clusters – evident from the two clustering plots we created earlier.

A commonly used measure to evaluate the quality of clusters is purity, which measures the extent to which the clusters contain a single class. Purity requires the notion of the class of a cluster. To compute purity, we assume that each cluster represents the class that is most frequent among its constituent data points. This is used to compute the accuracy by finding the ratio of the number of correctly assigned data points.

Clustering for Image Segmentation

As discussed in a previous chapter, an image can be seen as a multidimensional dataset where each pixel is represented by a set of values. K-means algorithm can be used to group all the pixels of an image into a predetermined number of clusters.

In this example, let's see how K-means can be used to cluster or segment various parts of an image. We'll download an image from Wikimedia Commons[1] and use it to discover how the K-means clustering method can help segment various parts of the image. We will use OpenCV implementation. You might want to install OpenCV for Python using the following:

```
%pip install opencv-python
```

We can now load the image in Python using opencv using the following:

```
import numpy as np
import cv2
import matplotlib.pyplot as plt

original_image = cv2.imread("C:\\Data\\Wikimedia Images\\port_
campbell.jpg")
plt.figure(figsize=(10,10))
plt.imshow(original_image)
```

The results would be visualized as shown in Figure 11-9. We wish to change the color space of the image and transform it into a shape that K-means can process. We will use cv2.cvtColor() to convert the image, followed by reshape.

---

[1] https://commons.wikimedia.org/wiki/File:Peterborough_(AU),_Port_Campbell_National_
Park,_Worm_Bay_--_2019_--_0863.jpg

Figure 11-9. *Image displayed in Python using imshow( )*

```
image = cv2.cvtColor(original_image, cv2.COLOR_BGR2RGB)
pixel_values = image.reshape((-1, 3))
pixel_values = np.float32(pixel_values)
```

OpenCV provides its implementation of K-means that requires the image pixels as samples, number of cluster, and a termination criterion that specifies when the algorithm should stop.

```
criteria = (cv2.TERM_CRITERIA_EPS + cv2.TERM_CRITERIA_MAX_ITER, 100, 0.2)
```

Now we can generate the clusters using cv2.kmeans(). We'd advice you to change the values of K and observe how the output changes.

```
K=5
_, labels, (centers) = cv2.kmeans(pixel_values, K, None, criteria, 10,
cv2.KMEANS_RANDOM_CENTERS)
```

This will assign a cluster (out of five) to each pixel in the picture. Centers will be an array containing five rows and three columns representing the average color of the cluster. Labels contain the cluster label (a number from 0 to 5) representing which cluster each pixel in the image has been assigned to.

We will make minor changes so that the cluster center values are converted to integers that can be represented as RGB colors.

```
centers = np.uint8(centers)
labels = labels.flatten()
```

We can now plot the segmented image.

```
segmented_image = centers[labels.flatten()]
segmented_image = segmented_image.reshape(image.shape)
plt.figure(figsize=(10,10))
plt.imshow(segmented_image)
plt.show()
```

Figure 11-10. *Five segments represented by different colors*

The image in Figure 11-10 shows the five clusters that each pixel has been assigned to in terms of its color. K-means is a very simple yet highly effective and popular technique for image segmentation.

Clustering Using DBSCAN

There are several other types of clustering algorithms that are more helpful in cases where the clusters are not expected to be spread evenly in a spherical fashion. To find such clusters with noticeable nonspherical shapes, we can use density-based methods that model clusters as dense regions separated by sparse regions.

DBSCAN, or Density-Based Spatial Clustering of Applications with Noise, is such a method that tries to combine the regions of space with density of data points greater than a predetermined threshold.

DBSCAN requires a predetermined distance, referred to as eps (for epsilon), which denotes the distance between points to be considered as part of a cluster, and min_ samples, which is the number of data points that must be present in the eps distance to form a dense cluster region. These parameters help control the tolerance of the algorithm toward noise and shape of the distribution that should be considered as a cluster.

DBSCAN iteratively expands the cluster by locating points that are presented in the regions of expected density. Figure 11-11 shows the expansion from three points within a definite region in space.

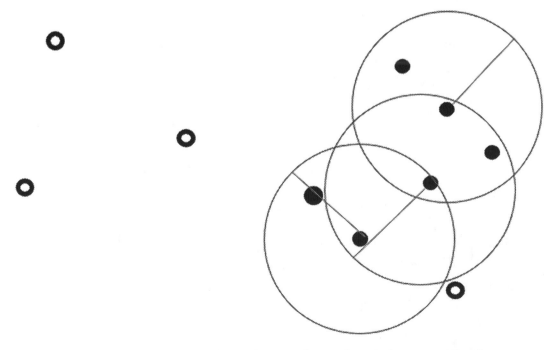

Figure 11-11. *DBSCAN expanding clusters through points present in close proximity*

For experiment in Scikit-learn, let's construct synthetic data that follows an arbitrary pattern.

```
from sklearn.datasets import make_moons, make_circles
X, y = make_moons(n_samples=1000, noise=0.1)
plt.scatter(X[:,0], X[:,1], edgecolor='k')
```

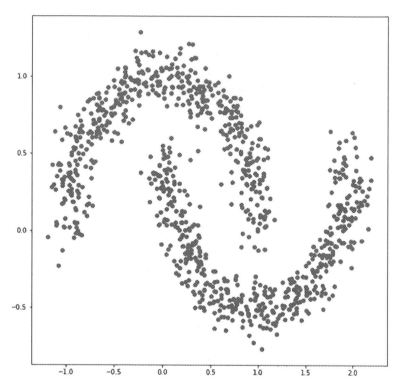

Figure 11-12. *Data points present in meaningful clusters which might not be captured by some algorithms*

The visualization in Figure 11-12 shows that there are intuitive clusters present in the dataset. If we try to locate the clusters using k-means, we might now achieve the clustering that we expect.

```
from sklearn.cluster import KMeans
kmeans = KMeans(n_clusters=2)
y= kmeans.fit_predict(X)
plt.scatter(X[:,0], X[:,1], c=y)
```

To find the clusters using DBSCAN, we can import DBSCAN from sklearn.clustering.

```
from sklearn.cluster import DBSCAN
dbscan = DBSCAN(eps=0.1, min_samples=2)
y= dbscan.fit_predict(X)
plt.scatter(X[:,0], X[:,1], c=y)
```

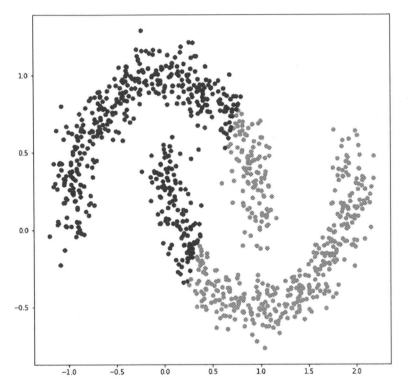

Figure 11-13. *Clusters detected by K-means*

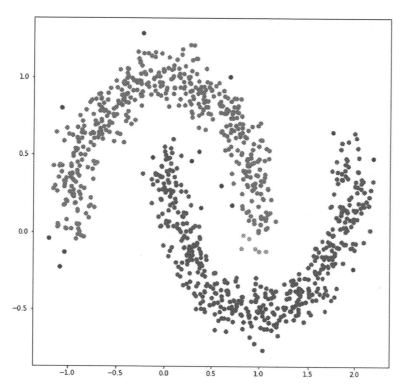

Figure 11-14. *Clusters detected by DBSCAN. Some points are detected as outliers*

DBSCAN is highly helpful in finding such clusters. It is robust and can be used to detect noise as well. The clusters present in Figure 11-13 and Figure 11-14 show a clear contract in terms of the quality of clusters that are generated by the K-means vs. the density-based algorithm.

Frequent Pattern Mining

Frequent pattern mining (FP mining) is a very popular problem that finds applications in retail and ecommerce. FP mining tries to discover repeating (frequent) patterns in the data based on combinations of items that occur together in the dataset. If you have dataset from transactions of a retail store or a supermarket, you can find patterns of the items that are commonly bought together, for example, milk and bread. Each transaction may contain one or more items that are purchased. FP mining algorithms provide us a way to find the most repeated combinations of such items that belong to a transaction.

Market Basket Analysis

A common term that is used in this context is market basket analysis. It is the process of discovering the items that are usually bought together. In the context of retail management, its application can help the decision makers in identifying where the product should be placed. The items that are often bought together can be placed close to each other, and in some deliberate cases, far away from each other, so that the customer has to walk through other shelves, which may make them purchase more than they wanted to.

Market basket analysis is also used to up-sell or cross-sell. This can help in identifying the retail opportunity by offering the customer more than they wanted to purchase, based on the buying behavior of customers in the past.

Such algorithms operate on the basis of finding association rules, that is, sets of items that are often bought together, along with a number indicating the support and a level of confidence on the rule.

```
Shampoo => conditioner [support = 2%, confidence =60%]
```

Support of 2% means that 2% of all the transactions show that the shampoo and the conditioner are purchased together. Confidence of 60% means that 60% of the customers who purchased the shampoo also bought the conditioner.

Association rules are considered interesting if they satisfy both a minimum support threshold and a minimum confidence threshold. These are set either by domain experts or through a set of iterations of careful analysis of the results.

Apriori algorithm is an iterative algorithm that grows a list of itemsets. It works on the idea that all the subsets of a frequent itemset must also be frequent. This is called apriori property. This means if there is a less frequent itemset, say, [Shampoo, Cockroach Spray], it doesn't satisfy the support threshold. If an item is added to the itemset resulting to [Shampoo, Cockroach Spray, Bread], it doesn't occur more frequently than the previous itemset.

Apriori algorithm begins by determining the support of itemsets in the transactional database. All the transactions with higher support value than the minimum or selected support value are considered. This is followed by finding all the rules of the subsets that have higher confidence value than the threshold or minimum confidence.

Apriori algorithm is computationally expensive. In most practical cases, the inventory may contain thousands (or larger order) of items. Another common issue is when support threshold is lowered to detect certain associations, it also increases the number of nonmeaningful associations.

Another algorithm called FP growth is implemented in most packages used to mine frequent itemsets. It first compresses the database representing frequent items into a tree-like structure, which retains the itemset association information. It then divides the compressed dataset into a set of conditional databases, each associated with one frequent item. Such databases are then mined separately. FP growth algorithm is much faster compared to apriori algorithm for larger datasets.

Frequent Pattern Mining in Python

Mlxtend (machine learning extensions) is a Python library that adds additional utilities and extensions to Python's scientific computing stack. For this experiment, you'll need to install mlxtend using pip install.

```
%pip install mlxtend
```

```
import pandas as pd
import numpy as np
```

```
import mlxtend
```

We've created a CSV containing items purchased in each transaction.

```
0,1,2,3,4,5,6
Bread,Wine,Eggs,Meat,Cheese,Butter,Diaper
Bread,Cheese,Meat,Diaper,Wine,Milk,Butter
Cheese,Meat,Eggs,Milk,Wine,,
Cheese,Meat,Eggs,Milk,Wine,,
```

The dataset we used in this code contains around 315 transactions, one transaction in each row. We can load the CSV and identify the unique items present across all the transactions.

```
df = pd.read_csv("fpdata.csv")

items = set()
for i in df:
    items.update(df[''+str(i)+''].unique())
items.remove(np.nan)

import mlxtend
from mlxtend.frequent_patterns import apriori, association_rules
```

We will now convert the dataset into a one-hot encoding form as we saw in previous chapters. Here's an alternate method to do the same. The encoded dataframe would look similar to the screenshot shown in Figure 11-15.

```
item_list = sorted(items)
encoded_vals = []

i=0
for index, row in df.iterrows():
    labels = dict()
    uncommons = list(items - set(row))
    commons = set(row).intersection(items)
    for item in commons:
        labels[item] = 1
    for item in uncommons:
        labels[item] = 0
    encoded_vals.append(labels)

one_hot_encoded_dataframe = pd.DataFrame(encoded_vals)
one_hot_encoded_dataframe
```

	Diaper	Meat	Bread	Butter	Wine	Eggs	Cheese	Milk	Donuts
0	1	1	1	1	1	1	1	0	0
1	1	1	1	1	1	0	1	1	0
2	0	1	0	0	1	1	1	1	0
3	0	1	0	0	1	1	1	1	0

Figure 11-15. *One-hot encoded dataframe, as expected by apriori algorithm implementation*

This format is expected by the implementation for apriori algorithm and association rules present in the mlxtend.frequent_patterns. We can generate a list of

```
from mlxtend.frequent_patterns import apriori, association_rules

freq_items = apriori(one_hot_encoded_dataframe, min_support = 0.2,
use_colnames=True)
freq_items['sup_count'] = freq_items['support']*315
freq_items
```

This should print a list of all the itemsets found that had a minimum support of 20% of the dataset. A truncated screenshot is shown in Figure 11-16. The results begin with one-item itemsets, which are not much meaningful because these are the frequencies of these items in the dataset. However, beyond row number 9, you can see two itemsets that show interesting buying patterns like bread and cheese. Later rows show three itemsets that are more meaningful. You can adjust the min_support to see more itemsets.

	support	itemsets	sup_count
0	0.406349	(Diaper)	128.0
1	0.476190	(Meat)	150.0
2	0.504762	(Bread)	159.0
3	0.361905	(Butter)	114.0
4	0.438095	(Wine)	138.0
5	0.438095	(Eggs)	138.0
6	0.501587	(Cheese)	158.0
7	0.501587	(Milk)	158.0
8	0.425397	(Donuts)	134.0
9	0.231746	(Bread, Diaper)	73.0
10	0.234921	(Wine, Diaper)	74.0
11	0.200000	(Diaper, Cheese)	63.0
12	0.206349	(Bread, Meat)	65.0
13	0.250794	(Wine, Meat)	79.0
14	0.266667	(Meat, Eggs)	84.0
15	0.323810	(Meat, Cheese)	102.0

16	0.244444	(Milk, Meat)	77.0
17	0.200000	(Bread, Butter)	63.0
18	0.244444	(Wine, Bread)	77.0
19	0.238095	(Bread, Cheese)	75.0
20	0.279365	(Milk, Bread)	88.0
21	0.279365	(Bread, Donuts)	88.0
22	0.200000	(Wine, Butter)	63.0
23	0.200000	(Butter, Cheese)	63.0
24	0.241270	(Wine, Eggs)	76.0
25	0.269841	(Wine, Cheese)	85.0
26	0.219048	(Milk, Wine)	69.0
27	0.298413	(Eggs, Cheese)	94.0
28	0.244444	(Milk, Eggs)	77.0
29	0.304762	(Milk, Cheese)	96.0
30	0.225397	(Milk, Donuts)	71.0
31	0.215873	(Meat, Cheese, Eggs)	68.0
32	0.203175	(Milk, Meat, Cheese)	64.0

Figure 11-16. *Frequent itemsets and support count detected by the apriori algorithm*

We can also print a list of association rules and the associated confidence scores.

```
rules = association_rules(freq_items, metric="confidence", min_
threshold=0.5)
rules
```

This shows interesting rules like those shown in Figure 11-17.

27	{Cheese}	{Milk}	0.501587	0.501587	0.304762	0.607595	1.211344	0.053172	1.270148
28	{Donuts}	{Milk}	0.425397	0.501587	0.225397	0.529851	1.056348	0.012023	1.060116
29	{Meat, Cheese}	{Eggs}	0.323810	0.438095	0.215873	0.666667	1.521739	0.074014	1.685714
30	{Meat, Eggs}	{Cheese}	0.266667	0.501587	0.215873	0.809524	1.613924	0.082116	2.616667
31	{Eggs, Cheese}	{Meat}	0.298413	0.476190	0.215873	0.723404	1.519149	0.073772	1.893773

Figure 11-17. *Association rules detected by the apriori algorithm*

Here, row number 29 indicates that those who bought meat and cheese also bought eggs, which is represented by support of meat and cheese as 0.32, support of eggs as 0.43, and support of the rule with these three items as 0.216. The associated confidence level is 0.667, which is quite high and reliable. You can explore other details in your dataset.

Summary

We have learned several unsupervised learning methods and discussed the situations where they are applied. With these tools and techniques, you can work on analysis of realistic datasets. In a later section, we will work on an end-to-end machine learning project using these techniques.

In the next chapters, we will discuss another set of methods that is getting highly popular in the past decade – neural networks and deep learning.

SECTION 3

Neural Networks and Deep Learning

CHAPTER 12

Neural Network and PyTorch Basics

In the past chapters, we have formed a foundation for machine learning approaches. These "traditional" machine learning methods have been used in academic research and industry applications for decades. However, the subject of focus in the new innovations in the past few years has been neural networks – the capability, the performance, and the versatility of various deep neural network architectures.

This and the next few chapters primarily focus on neural networks as well as deep neural network architectures, mainly the convolutional neural networks and the recurrent neural networks that are directly applicable in many situations. In this chapter, we will discuss how neural networks work, how are they applicable in so many different types of solutions, and how PyTorch can be used.

There are several software libraries and toolkits that have become popular in the past few years. Python has become the most popular choice for most of the projects that involve machine learning – and for deep learning, PyTorch is one of the competing tools whose popularity has been increasing in the recent past. We will limit ourselves to this library – though the API and the way of using the library might be different compared to the other tools, the ideas are still directly relevant and applicable.

In this chapter, we will begin with basics of perceptrons, which are the basic building block of neural networks. This will also involve basic mathematical operations that are required in constructing neural networks. We will then have a crisp introduction of PyTorch and look at the basic features. We will learn to do basic computations using PyTorch. Throughout this chapter and the next, you will be introduced to features of PyTorch based on their relevance.

© Ashwin Pajankar and Aditya Joshi 2022
A. Pajankar and A. Joshi, *Hands-on Machine Learning with Python*, https://doi.org/10.1007/978-1-4842-7921-2_12

Neural networks are interconnected nodes of computations that are at the heart of deep learning algorithms. The most basic element of neural networks is called a perceptron, which performs very basic vector arithmetic operations. Perceptrons can be combined together to depend on each other's results for further computation and thus be arranged in layers of computing units. Such networks are called neural networks.

We will discuss more details about neural networks, starting with the basic unit, the perceptron, in the further sections.

Though the simplicity of design is the main source of power and popularity of neural networks, these computations can often grow too big and complex to program and manipulate using basic programming tools, which led to the rise of frameworks for neural network programming.

PyTorch is one of the most popular tools often applauded for being simple and more Pythonic – thus leading to easy learning and improved developer productivity. PyTorch is also at par, and, in some cases, faster than other popular deep learning libraries. The benefits are summarized by a highly cited AI scientist in his tweet[1] (Figure 12-1).

Figure 12-1. *Andrej Karpathy's tweet about PyTorch. Andrej is presently leading AI and autopilot vision at Tesla*

Installing PyTorch

One of the most preferred ways to install PyTorch is to use Anaconda distribution's package manager tool – conda. If you do not have Anaconda installed, you may download the suitable package for your system at the Anaconda website.[2] Anaconda

---

[1] https://twitter.com/karpathy/status/868178954032513024

[2] www.anaconda.com/products/individual

is a popular Python distribution that gives powerful environment management and package management capabilities. Once you have installed Anaconda, you can use this command:

```
conda install pytorch torchvision torchaudio cudatoolkit=10.2 -c pytorch
```

This will instruct conda to install PyTorch and the required libraries, including cudatoolkit, which provides an environment for creating high-performance gpu-based programming.

Alternatively, you can install using Python's package manager, pip.

```
pip3 install torch torchvision torchaudio
```

To explore more alternative ways to install, you can refer to the Getting Started[3] page of PyTorch, which gives you more options to configure your installation depending on your system and requirements.

After installation, you can perform a simple test to verify that your PyTorch installation works as expected. Open Jupyter Notebook or Python Shell to import torch and verify the PyTorch version.

```
import torch
torch.__version__
Out: '1.9.1'
```

PyTorch Basics

In PyTorch, tensor is a term that refers to any generic multidimensional array. In practice, tensor establishes a multilinear relationship between the sets of algebraic objects related to a vector space. In PyTorch, tensor is the primary data structure that encodes the inputs, outputs, as well as the model parameters.

Creating a Tensor

Tensors are similar to Ndarrays in NumPy. You can create a tensor from a Python's list or multidimensional lists of lists. You can also create a tensor from an existing NumPy array.

---

[3] https://pytorch.org/get-started/locally/

```
mylist = [1,2,3,4]
mytensor = torch.tensor(mylist)
mytensor
Out: tensor([1, 2, 3, 4])
```

As you might expect, you can also create a tensor from a NumPy array.

```
import numpy as np
myarr = np.array([[1,2],[3,4]])
mytensor_2 = torch.from_numpy(myarr)
mytensor_2
Out: tensor([[1, 2],
        [3, 4]], dtype=torch.int32)
```

When you create a tensor from a NumPy array, it is not copied to a new memory location – but both the array and tensor share the same memory location. If you make any change in the tensor, it will be reflected in the original array from which it was created.

```
mytensor_2[1,1]=5
myarr
Out: tensor([[1, 2],
        [3, 5]], dtype=torch.int32)
```

Inversely, you can also use `mytensor_2.numpy()` to return a NumPy array object that shares the same data. Just like NumPy Ndarrays, PyTorch tensors are also homogeneous; that is, all the elements in the tensor have same data type. There are other tensor creation methods similar to NumPy's array creation methods. This is an example of creating a simple tensor.

```
torch.zeros((2,3))
Out: tensor([[0., 0., 0.],
        [0., 0., 0.]])
```

This will create a tensor of shape 3x3 with all values as zeros. A similar function in NumPy is np.zeros(3,3). It returns an array of shape 3x3. Though the representation is similar, tensors are the primary unit of data representation in PyTorch. You can use a similar functions to create arrays of ones, or random values of user-defined size.

```
torch.ones((2,3))
Out: tensor([[1., 1., 1.],
        [1., 1., 1.]])
```

```
torch.rand((2,3))
Out: tensor([[0.0279, 0.5261, 0.9984],
        [0.7442, 0.3559, 0.3686]])
```

PyTorch also includes a method to create or initialize a tensor with properties (like shape) of another tensor.

```
torch.ones_like(mytensor_2)
Out: tensor([[1, 1],
        [1, 1]], dtype=torch.int32)
```

Tensor Operations

PyTorch tensors support several operations in a similar manner as NumPy's arrays – though the capabilities are more. Arithmetic operations with scalers are broadcasted, that is, applied to all the elements of the tensor. Matrix operations between tensors of compatible shapes are applied in a similar fashion.

```
myarr = np.array([[1.0,2.0],[3.0,4.0]])
tensor1 = torch.from_numpy(myarr)
```

```
tensor1+1
Out: tensor([[2., 3.],
        [4., 5.]], dtype=torch.float64)
```

```
tensor1/ tensor1
Out: tensor([[1., 1.],
        [1., 1.]], dtype=torch.float64)
```

```
tensor1.sin()
Out: tensor([[ 0.8415,  0.9093],
        [ 0.1411, -0.7568]], dtype=torch.float64)
```

```
tensor1.cos()
Out: tensor([[ 0.5403, -0.4161],
        [-0.9900, -0.6536]], dtype=torch.float64)
```

```
tensor1.sqrt()
Out: tensor([[1.0000, 1.4142],
        [1.7321, 2.0000]], dtype=torch.float64)
```

The functions for describing the data can also be used in a similar manner:

```
mean, median, min_val, max_val = tensor1.mean(), tensor1.median(), tensor1.
min(), tensor1.max()
```

```
print ("Statistical Quantities: ")
print ("Mean: {}, \nMedian: {}, \nMinimum: {}, \nMaximum: {}".format(mean,
median, min_val, max_val))
print ("The 90-quantile is present at {}".format(tensor1.quantile(0.5)))
```

These operations give the following output:

```
Statistical Quantities:
Mean: 2.5,
Median: 2.0,
Minimum: 1.0,
Maximum: 4.0
The 90-quantile is present at 2.5
```

Similar to NumPy, PyTorch also provides operations like cat, hstack, vstack, etc., to join the tensors. Here are the examples:

```
tensor2 = torch.tensor([[5,6],[7,8]])
torch.cat([tensor1, tensor2], 0)
```

This method would concatenate the two tensors. The direction (or axis) of concatenation is provided as the second argument. 0 signifies that the tensors will be joined vertically, and 1 would signify that they will be joined horizontally.

```
tensor([[1., 2.],
        [3., 4.],
        [5., 6.],
        [7., 8.]], dtype=torch.float64)
```

Other similar functions are hstack and vstack, which can also be used to join two or more tensors horizontally or vertically.

```
torch.hstack((tensor1,tensor2))
Out: tensor([[1., 2., 5., 6.],
        [3., 4., 7., 8.]], dtype=torch.float64)
torch.vstack((tensor1,tensor2))
Out: tensor([[1., 2.],
        [3., 4.],
        [5., 6.],
        [7., 8.]], dtype=torch.float64)
```

There's a reshape function that changes the shape of the tensor. To convert the tensor into a single row with an arbitrary number of columns, we can use shape as (1,-1):

```
torch.reshape(tensor1, (1, -1))
Out: tensor([[1., 2., 3., 4.]], dtype=torch.float64)
```

We will continue discussing more operations based on their use in the future sections and chapters.

Perceptron

Perceptron, shown in Figure 12-2, is the simplest form of neural network. It takes one or more quantities often describing features of a data item as input, performs a simple computation on it, and produces a single output. The simplest form of perceptron is a single-layer perceptron – it is easy to understand, quick to run, and simple to implement, though it can classify only linearly separable data.

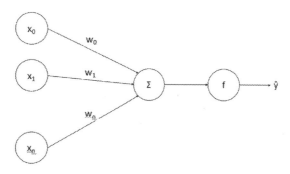

Figure 12-2. *A simple perceptron structure with constituent computations*

As illustrated in Figure 12-2, perceptron applies a simple computation to predict the class label based on the input vector x, with features being represented as $[x_1, x_2, x_3 \ldots x_n]$, and a weight vector, with features being represented as $[w_1, w_2, w_3, \ldots w]$. We usually add an additional bias term that doesn't affect the output based on the input as w_0. To ease the computation, we add an input feature as x_0, with its value being set as 1. Thus, the two vectors, x and w, lead to the final output based on a simple step function:

$$\hat{y} = f(x) = \begin{cases} 1, & x^t w > 0 \\ 0, & otherwise \end{cases}$$

Here, x is the input vector with n+1 dimensions, w is the weight vector, and the objective of the learning process is to learn the best possible weights so that the error computed by comparing the results with the training labels is minimized.

To train the perceptron, we first initialize the weights with random values. We use the training dataset to find the predicted output based on the formula shown previously. Because the algorithm has not learned the right set of weights yet, the results may be far than we expected – thus leading to an error. To reduce the noise in next iterations, we apply the following update to the weights based on the current outputs:

$$w = w + \alpha(y - \hat{y}).x$$

Here, we have also added a step parameter, α, which controls how severely are the weights impacted. We iteratively repeat this process till a predetermined number of times (or till convergence) and hope to reach a good enough weight vector by the end, which may lead to a low error. To predict the output, we simply plug in the features x to the same computation.

The computation function we saw in the previous section is called a stepper function. In many cases, you might rather see sigmoid function, that is:

$$y = \frac{1}{1 + e^{-w.x}}$$

Let's see how to program these first using basic Python and NumPy, and later we will use PyTorch to do the same.

Perceptron in Python

We will first create a simple separable dataset using Scikit-learn's dataset module. You can use any other dataset that we have used before.

```
from sklearn import datasets
import matplotlib.pyplot as plt
X, y = datasets.make_blobs(n_samples=100,n_features=2, centers=2, random_
state=42, shuffle=1)
```

These lines will create 100 rows of data with two features, which is divided into two major blobs. Let's visualize it before building the perceptron.

```
fig = plt.figure(figsize=(10,8))
plt.plot(X[:, 0][y == 0], X[:, 1][y == 0], 'b+')
plt.plot(X[:, 0][y == 1], X[:, 1][y == 1], 'ro')
plt.xlabel("Feature 1")
plt.ylabel("Feature 2")
```

The data points are clearly visible as shown in Figure 12-3, and we deliberately chose this kind of randomly generated dataset to keep the classification boundary simple.

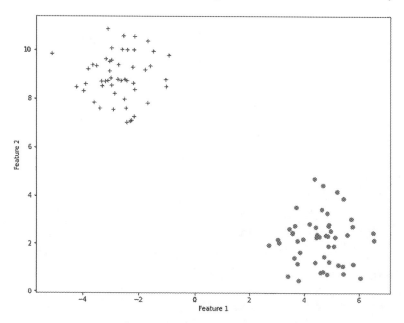

Figure 12-3. *Simple dataset generated using make_blobs*

We have X and y. Let's create fit() and predict() methods from scratch. You know that weights are updated based on the predicted value of each data point at the current step. In the formula $w + \alpha(y - \hat{y}).x$, we already know x and y; α is a hyperparameter that you can configure to set the steps, and w is the set of weights that we will learn in this process.

Because our dataset contains two features, we need three (2 + 1 for the bias term) weights in the weight vector. Let's implement the predict function first.

```python
def predict(X, weight):
  return np.where(np.dot(X, weight) > 0.0, 1, 0)
```

This implements the formula for $\hat{y}$ by first computing the product between the input data and weight and then applies the case to compare if the resulting product is larger than 0. This function can act as the prediction function as well as help us find the right value to update the weights.

Remember there are three weights, though the dataset still has two columns. We will add a column to compensate for the bias, thus leading to a shape of 100x3. We will initialize the weights to random values.

```python
X = np.concatenate( (np.ones((X.shape[0],1)), X), axis=1)
weight = np.random.random(X.shape[1])
```

After initialization, we will run an iterative process till a predetermined number of iterations (or epochs) and, within each iteration, process each point and update the weights. The fit() method should now look like the following:

```python
def fit(X, y, niter=100, alpha=0.1):
  X = np.concatenate( (np.ones((X.shape[0],1)), X), axis=1)
  weight = np.random.random(X.shape[1])

  for i in range(niter):
    err = 0
    for xi, target in zip(X, y):
        weight += alpha * (target - predict(xi, weight)) * xi
  return weight
```

We have not structured the code as a class which might store the weights internally – but have to return it, which can be supplied to the predict method. To learn the weights, we can now call

```
w = fit(X,y)
w
Out: array([ 0.21313539,  0.96752865, -0.84990543])
```

W is a weight vector with three values representing the bias and corresponding coefficient for each feature. We can use the predict() method to predict the output. Let's pick some random elements from X to compare how our perceptron labels them:

```
random_elements = np.random.choice(X.shape[0], size=5, replace=False)
X_test = X[random_elements, :]
```

X_test will now contain five random rows from the dataset. Before calling the predict method, we will need to add an additional column with ones.

```
X_test = np.concatenate( (np.ones((X_test.shape[0],1)), X_test), axis=1)
```

Now let's call the predict method and compare the results with the actual values.

```
print (predict(X_test, w))
print (y[random_elements])
Out:
[0 0 1 0 0]
[0 0 1 0 0]
```

The results look good because the dataset is too simple – however, remember that a simple perceptron doesn't perform well in cases where decision boundaries are not so clear. We will discuss how to combine such simple computation units to create a more complex neural network in the next chapters.

Artificial Neural Networks

A simple perceptron learns how important each feature of the dataset is through a single threshold logic unit, when it attempts to combine the weighted sum of features and pass it to a function. We used a simple step function that was implemented by if-else conditions in Python, or np.where function while using NumPy. We may use a sigmoid function or other alternate functions to manipulate how and when a feature set produces the output – or activates the neuron. We can combine multiple activations like these and connect them in the form of a fully connected layer.

Such networks surpass the simplicity of using a simple perceptron and allow you to create multiple fully connected layers, thus leading to creation of hidden layers, thus creating a multilayer perceptron as shown in Figure 12-4. The output may become input to the next layer, which is further manipulated by a new set of weights at the next layer.

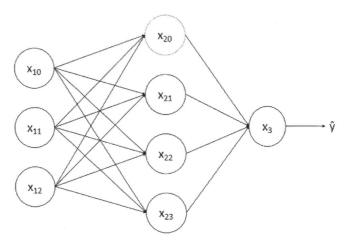

Figure 12-4. *A simple multilayer perceptron with one output unit*

Computation units can be arranged in a different fashion to create more diverse deep neural networks. The convolutional neural networks (CNNs) use a special kind of layers that apply a filter to an input image by sliding the filter to produce an activation map. CNNs are a highly preferable choice in many computer vision (CV) or natural language processing (NLP) applications. We will discuss more about them in depth in Chapter 14.

Another popular neural network architecture is called recurrent neural network (RNN), which has an internal state whose value is maintained based on input data in order to produce an output using a combination of the state and an input sample. This might also update the internal state and affect the future outputs. This thus helps interpret the sequential information in the data, which is highly helpful in NLP applications. We will study about this in Chapter 15.

Summary

In this chapter, we have discussed the basics of neural networks and have started to explore PyTorch to define tensors and create simple neural units that can learn to classify data. The next chapter discusses the algorithms that are used to learn network weights in neural networks, thus leading to feedforward and backward propagation.

CHAPTER 13

Feedforward Neural Networks

Artificial neural networks (ANN)s are collections of interconnected computation units modelled based on the neurons in the brain so that the program thus created is capable of learning patterns in structured, textual, speech, or visual data. The basic computational unit in an artificial neural network is thought to be similar to (or rather, inspired from) a neural cell that accepts input signals from multiple sources, operates on them, and activates based on the given condition, which passes the signal to other neurons connected to it. Figure 13-1 shows the symbolic link between the biological neuron and the artificial neuron.

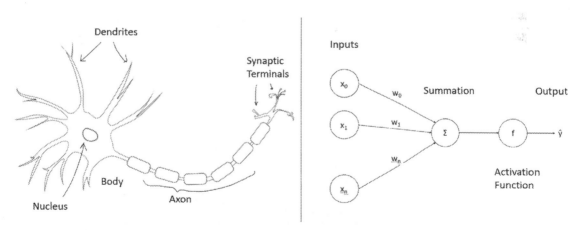

Figure 13-1. *Comparison of a biological neuron with an artificial neuron*

Just like neurons come together to form parts of the brain that are together responsible for recognizing some patterns or perform an action, artificial neural networks consist of a large number of such units. As we saw in the previous chapter, the neurons or computational units combine signals from multiple sources, apply an

© Ashwin Pajankar and Aditya Joshi 2022
A. Pajankar and A. Joshi, *Hands-on Machine Learning with Python*, https://doi.org/10.1007/978-1-4842-7921-2_13

activation function, and pass the processed signal to other neurons connected to it. In a practical application, there can be a dozen to millions of neurons present that can be trained to operate and activate based on the input and expected output values in training data.

In this chapter, we will study neural networks in which layers of computational units are tied to each other till an output layer, in which the input signals are operated in each layer and fed forward, while the training process compares the output and traces back the changes required to develop a better neural network.

The concepts covered in this chapter form the basis of more advanced architectures of deep neural networks. We will discuss the process of training neural networks through a process called backpropagation, which is based on gradient descent. We will use PyTorch to create neural networks for regression and classification problems.

Feedforward Neural Network

Feedforward neural network is a simple form of artificial neural network in which the computation units pass the values gradually toward the output, combining them in a highly efficient manner to lead to improved results. The computation units in a feedforward neural network propagate the information in the forward direction (there are no cycles or backward links), through the hidden nodes and then to the output nodes. A simple example of a feedforward neural network is shown in Figure 13-2.

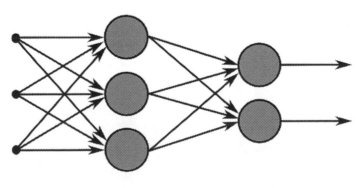

hidden layer   output layer

Figure 13-2. A simple feedforward neural network

Training Neural Networks

Backpropagation is the process for training neural networks. Backpropagation is based loosely on techniques that are being used since the 1960s, though this was thoroughly defined in 1986 by Rumelhart, Hinton, and Williams, which was followed by Yann LeCun's work in 1987. During this time period, there were several promising works on neural networks, which form the basis of the field of deep learning today, though they couldn't catch much attention in the general public due to limited computational infrastructure of that time. Later, around the 2010s, the cost of computer processors and graphics processors (GPUs) declined sharply, thus giving rise to the refinement of decade-old models and the creation of novel neural network architectures, leading to their use in speech recognition, computer vision, and natural language processing.

Gradient Descent

Training of neural networks requires a process called gradient descent, an iterative algorithm that is used to find the minimum (or maximum) value of a loss or a cost function. Imagine a regression problem (similar to what we saw in Chapter 7) in which a continuous output variable is determined based on a continuous input variable. In most practical cases, the predicted output, shown as a line in Figure 13-3, will not be exactly the same as the actual (expected) output. This difference is called errors, or residuals, and the learning algorithm aims to minimize the total residuals or some other aggregation of residuals.

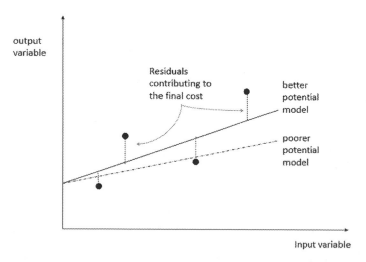

Figure 13-3. *Errors in a regression model*

The learning process tends to learn the parameters of equation of a line in the form of $y = w_0 + w_1x_1 + w_2x_2 + \cdots$. To simplify our example, we will stick to only one variable, thus leading to $y = w_0 + w_1x_1$. One of the approaches used to solve such problems is gradient descent. In this example, we define an optimization function, here, a cost function, which shows how far the model's results are with respect to the actual values in the training data. In linear regression, we can use mean squared error, which is the average of squares of differences in the values predicted by the model and the actual values.

$$J = \frac{1}{n}\sum_{i=1}^{n}\left(pred_i - y_i\right)^2$$

The idea behind gradient descent is that the well-trained model should be the one in which the cost function is minimized. Each possible set of slopes (w_0, w_1,...) will produce a different model. Figure 13-4 shows the change in cost with respect to a slope, say, w_1.

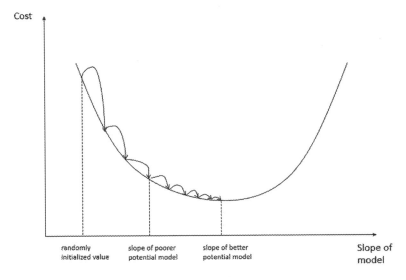

Figure 13-4. *Gradient descent algorithm aims to find the parameters that minimize the cost*

Our aim is to find a slope that produces the minimum cost. You can see the corresponding point at the lowest point of curve in the figure. Gradient descent algorithm begins with a randomly initialized value, and based on the slope of the cost at the point given by the partial derivative with respect to the slope, the algorithm changes:

$$\frac{\partial\left(cost\right)}{\partial m} = \frac{1}{n}\frac{\partial}{\partial m}\left(pred_i - y_i\right)^2$$

which resolves to

$$\frac{\partial\left(cost\right)}{\partial m}=\frac{-2}{n}x\left(pred_{i}-y_{i}\right)$$

This denotes the update in the value of m that should ideally lead toward a model with low cost. This process is the basis of backprop algorithm, and through the right choice of the loss function, we can train neural networks for much more complex problems.

If we translate this idea to neural network terminology, gradient descent provides us a way to update the weights of a one-layer neural network like the one we saw in the previous chapter.

Backpropagation

In a multilayer network, this method can be directly applied at the final layer where we can find the difference in the actual (or target) and the predicted value; however, this can't be applied in the hidden layers because we don't have any target values to compare. To continue updating the weights on all the individual cells of the neural network, we calculate the error in the final layer and propagate that error back from the last layer to the first layer. This process is called backpropagation.

Let's consider a simplified neural network with one hidden layer as shown in Figure 13-5. We have only three nodes; the first one represents the input, the second is a hidden layer that performs computation on the input based on weight w_1 and bias b_1, and the third is an output layer, which also performs computation on the output from the hidden layer based on weight w_1 and bias b_1.

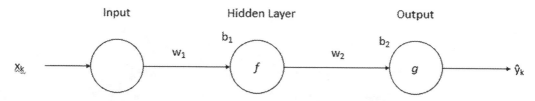

Figure 13-5. *Simplified neural network with one hidden layer*

Here, the first node accepts the input and forwards it to the second node, the hidden layer. Hidden layer applies the weight w_1 to the input and adds bias b_1, thus producing $w_1x_k+b_1$, which is then applied with the activation function f. Thus, the output of hidden layer is $f(w_1x_k+b_1)$.

The output of hidden layer is being forwarded as the input of the third unit. This will be multiplied by w_2 and added to bias b_2. Thus, the input of the third unit is w_2 $f(w_1x_k+b_1) + b_2$.

When the activation function g is applied to it, it produces the output $g(w_2 f(w_1x_k+b_1) + b_2)$, which is the predicted output, also denoted in Figure 13-5 as $\hat{y}_k$.

This process of forward propagation happens in each iteration during the training phase. Once we know the predicted value for all the items in the training dataset, we can find the loss or the cost function. For this explanation, let's continue the same loss function we defined in the previous section.

$$J = \frac{1}{n}\sum_{k=1}^{n}\left(\hat{y}_k - y_k\right)^2, \qquad\qquad \text{where}$$

$$\hat{y}_k = g\left(w_2 \, f\left(w_1 x_k + b_1\right) + b_2 \right)$$

We know that we can compute the derivative of loss function with respect to w_1 to update w_1 in order to reduce the overall loss in the next iteration.

$$\frac{\partial(J)}{\partial w_1} = \frac{1}{n}\sum_{k=1}^{n}2\left(\hat{y}_k - y_k\right)\frac{\partial\hat{y}_k}{\partial w_1}$$

This leads to another quantity that is resolved using chain rule as follows:

$$\frac{\partial\hat{y}_k}{\partial w_1} = g'\left(w_2 \, f\left(w_1 x_k + b_1\right) + b_2 \right)\left\{w_2 f'\left(w_1 x_k + b_1\right)\left(x_k\right)\right\}$$

Or

$$\frac{\partial\hat{y}_k}{\partial w_1} = w_2 x_k f'\left(w_1 x_k + b_1\right) g'\left(f\left(w_1 x_k + b_1\right) + b_2 \right)$$

We can find partial derivative with respect to w_2 as

$$\frac{\partial(J)}{\partial w_2} = \frac{1}{n}\sum_{k=1}^{n}2\left(\hat{y}_k - y_k\right)\frac{\partial\hat{y}_k}{\partial w_2}$$

$$\frac{\partial \hat{y}_k}{\partial w_2} = \{f(w_1 x_k + b_1)\} \, g'(\, w_2 \, f(w_1 x_k + b_1) + b_2\,)$$

Thus, in more complex networks, we can continue computing partial derivatives in the backward direction. We will notice that there are several quantities that we have computed previously, for example, $f(w_1 x_k + b_1)$ in the preceding example. You can see that in each iteration, the intermediate values and the output values are computed during a forward pass, followed by the process of updating weights using gradient descent starting from the output layer, moving backward till all the weights are updated. This is called backpropagation.

Loss Functions

In the previous explanation, we used a loss function called mean squared error (MSE). Due to its nature, this kind of loss function is suitable for regression problems where the output is a continuous variable. There are several other common loss functions that you can use depending on the problem in hand.

Mean Squared Error (MSE)

This averages the sum of squares of the error between actual value and predicted value. This penalizes the model for large errors and ignores small errors. This is also called L2 loss. For two values, y and $\hat{y}$, usually expected output and predicted output, the error component for each training sample is given by

$$loss(y, \hat{y}) = (y - \hat{y})^2$$

Mean Absolute Error

Instead of considering the squares, we can simply look at the absolute sum of squares and take a mean across the dataset. This is also called L1 loss, and it is robust to outliers.

$$loss(y, \hat{y}) = |y - \hat{y}|$$

Negative Log Likelihood Loss

In simple classification problems, negative log likelihood loss is an efficient option that encourages the models in which the prediction is made correctly with high probabilities and penalizes it when it predicts the correct class with smaller probabilities.

$$loss(y, \hat{y}) = -log(y)$$

Cross Entropy Loss

This is a suitable function to use in classification problems. It penalizes the model for producing wrong output with high probability. It is one of the mostly used loss functions when training a classification problem with C classes.

$$loss(y, \hat{y}) = -\sum y log \, \hat{y}$$

Hinge Loss

In problems where we want to learn nonlinear embeddings, hinge loss measures the loss given an input tensor x and a label tensor y (containing 1 or -1). This is usually used for measuring whether two inputs are similar or dissimilar.

$$loss(x, y) = \begin{cases} x, if \, y = 1 \\ \max\{0, \Delta - x\}, if \, y = -1 \end{cases}$$

For more loss functions that are defined in PyTorch, you can look at the official documentation.[1]

ANN for Regression

Let's use PyTorch to create a simple neural network for a regression problem. Let's begin with creating a simple dataset with one independent variable (X) and a dependent variable (y), where there might be a linear-like relationship between X and y. We will create tensors of shape [20,1], thus representing 20 inputs and 20 output values. The output plot is shown in Figure 13-6.

---

[1] https://pytorch.org/docs/stable/nn.html#loss-functions

```
from matplotlib import pyplot
import torch
import torch.nn as nn
x = torch.randn(20,1)
y = x + torch.randn(20,1)/2
pyplot.scatter(x,y)
pyplot.show()
```

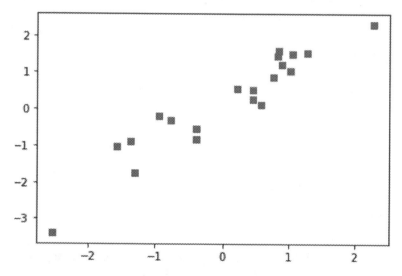

Figure 13-6. *Randomly generated samples for regression*

The data is ready. This model is a simple sequential model, which will have an input layer, followed by an activation function, and another output layer.

```
model = nn.Sequential(
    nn.Linear(1,1),
    nn.ReLU(),
    nn.Linear(1,1)
)
```

Activation function is the function f that is applied to the weighted inputs $f(w_i x_i)$. ReLU, or Rectified Linear Unit, is a simple function that produces an output of 0 for any negative inputs and produces the unchanged input value for positive inputs. We will discuss about ReLU and other activation functions in the next section.

Because we have only one input variable, we expect to learn two weights, w_1 and b. Due to ease of implementation, we refer to b as w_0. For the input linear layer and the output linear layer, there will be two sets of weights each, thus a total of four weights initialized randomly, that we need to learn during training. Let's see the model parameters to understand the quantities that will be learned.

```
list(model.parameters())
Out: [Parameter containing:
 tensor([[-0.7681]], requires_grad=True),
 Parameter containing:
 tensor([0.2275], requires_grad=True),
 Parameter containing:
 tensor([[0.1391]], requires_grad=True),
 Parameter containing:
 tensor([-0.1167], requires_grad=True)]
```

We can now begin the process to learn the weights that minimizes the loss function using the method defined as the optimizer. However, remember that PyTorch requires the data to be in the form of a tensor. Let's quickly (1) scale the data in the 0–1 range and (2) convert to tensor.

```
x = (x-x.min())/(x.max()-x.min())
y = (y-y.min())/(y.max()-y.min())
```

We now need to initialize mean squared error (MSE) loss function and an optimizer, which will use stochastic gradient descent for updating the weights.

```
lossfunction = nn.MSELoss()
optimizer = torch.optim.SGD(model.parameters(), lr=0.05)
```

While initializing the optimizer, we have defined a learning rate of 0.05. This affects how quickly (or slowly) the weights are updated.

The learning process requires multiple iterations in three steps:

1. Forward propagation: Using the current set of weights, compute the output.

2. Computation of losses: Compare outputs with the actual values.

3. Backpropagation: Use the losses for updating weights.

Here, we use a for loop to iterate over 50 epochs. In this process, we will also keep track of losses so that we can later visualize how errors change over epochs.

```
loss_history = []

for epoch in range(50):
    pred = model(x)
    loss = lossfunction(pred, y)

    loss_history.append(loss)

    optimizer.zero_grad()
    loss.backward()
    optimizer.step()
```

After 50 iterations, we expect the losses to be low enough to produce a decent result. Let's visualize how the losses changed by plotting a chart of loss_history. We must remember that the loss object produced by lossfunction() will also contain data as a tensor, and we need to detach it so that Matplotlib can process it.

```
import matplotlib.pyplot as plt
plt.plot([x.detach() for x in loss_history], 'o-', markerfacecolor='w',
linewidth=1)
plt.plot()
```

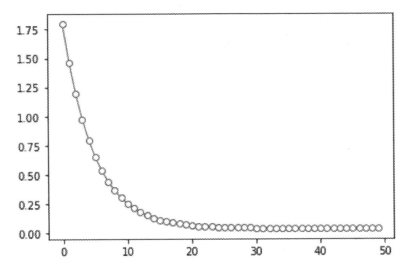

Figure 13-7. *Loss reduces as the model is trained for more epochs and converges after a point*

It is evident from Figure 13-7 that the losses decrease rapidly till the tenth epoch, after which the errors were so low that the gradient reduced, and the further changes were slower – till somewhere around the 30th epoch, after which the loss stayed the same and the change in weights was minimal.

Let's look at the results produced by the system. The result thus produced is shown in Figure 13-8.

```
predictions = model(x)
plt.plot(x, y, 'rx', label="Actual")
plt.plot(x, predictions.detach(), 'bo', label="Predictions")
plt.legend()
plt.show()
```

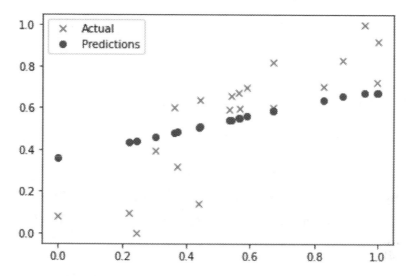

Figure 13-8. *Actual values and output predicted by a simple neural network*

This overly simple one-layer neural network might not always give best results for problems in which you can use fairly straightforward statistical solutions. It is possible that despite the graph showing evident reduction in losses, the final regression line might not be as closely fitting as you might expect. You can define the model and train it multiple times to see the difference due to random initialization.

With this model, you are now ready to work with more complex neural network architectures. We will first build a multilayer neural network with different activation function and use it to classify Iris flowers using the same dataset we have used so far.

Activation Functions

Each computation unit in a neural network accepts the input, multiplies weights, adds the bias, and applies an activation function to it before forwarding it to the next layer. This becomes input for the computation units in the next layer. In this example, we used Rectified Linear Unit (ReLU), which returns x, for an input x if x>0; otherwise, it returns 0. Thus, if a unit's weighted computation yields a negative value, ReLU will make it as 0, which will be the input to the next layer. If the input is zero or negative, the derivative of activation function is 0, otherwise, 1.

There are many activation functions that have been defined. Here are some of the activation functions that you might often see being used.

ReLU Activation Function

Rectified Linear Unit is a simple and efficient function that enables the input as it is, if it is positive and doesn't activate for the negative input; thus, it rectifies the incoming signal. It is computationally fast, nonlinear, and differentiable. It is defined as

$$ReLU(x) = \max(0,x)$$

However, because in case of negative inputs the neuron doesn't affect the output at all, its contribution to the output becomes zero and thus doesn't learn during backpropagation. This is solved by a variation of ReLU called **Leaky ReLU**.

Figure 13-9 shows the graph of ReLU and Leaky ReLU.

$$LeakyReLU(x) = max(0,x) + negative\_slope * min(0,x)$$

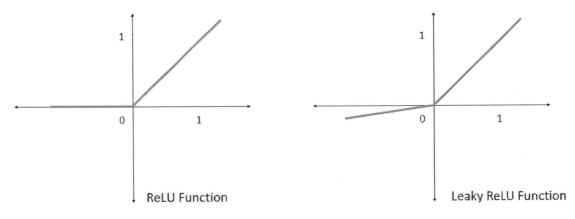

ReLU Function   **Leaky ReLU Function**

Figure 13-9. *ReLU and Leaky ReLU activation function*

Leaky ReLU produces a relatively small output for negative signals, which can be configured by changing the negative_slope.

Sigmoid Activation Function

Sigmoid function produces an output between 0 and 1, with output values close to 0 for a negative input and close to 1 for a positive input. The output is 0.5 for input of 0. You can see the graph for sigmoid activation function in Figure 13-10. This function is highly suitable for classification problems, and if used in the output layer, the output value that is between 0 and 1 can be interpreted as a probability.

$$Sigmoid(x) = \frac{1}{1 + exp(-x)}$$

However, sigmoid is computationally more expensive. If the input values are too high or too small, it can cause the neural network to stop learning. This problem is called vanishing gradient problem.

Tanh Activation Function

Tanh function is similar to sigmoid function but produces an output between -1 and 1, with output values close to -1 for a negative input and close to 1 for a positive input. The function crosses the origin at (0,0). The graph for tanh function is shown in Figure 13-10. You can see that though the two functions look similar, tanh function is significantly different.

$$tanh(x) = \frac{exp(x) - exp(-x)}{exp(x) + exp(-x)}$$

Despite the similar shape, the gradients of tanh function are much stronger than sigmoid function. It is also used for layers which we wish to pass the negative inputs as negative outputs.

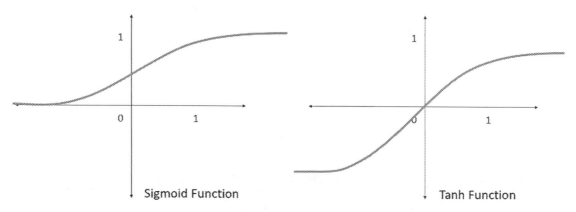

Figure 13-10. *Sigmoid and tanh activation function*

If a very negative input is provided to sigmoid function, the output value will be close to zero, and thus, the weights will be updated very slowly during backpropagation. Tanh thus improves the performance of the network in such situations.

Multilayer ANN

We can make the network slightly larger by (1) adding more computation units in the layers and (2) adding more layers. We can modify our network by editing how we specified the layers in the previous example.

```
model = nn.Sequential(
    nn.Linear(1,8),
    nn.ReLU(),
    nn.Linear(8,4),
    nn.Sigmoid(),
    nn.Linear(4,1),
)
```

Here, we have defined an input layer that takes one input and forwards it to the layer with eight units. This is followed by ReLU activation function before moving to a layer with eight units, which then forwards to another layer with four units. This is followed by sigmoid activation, which forwards to the output layer with one unit.

We can modify the output layer to contain more than one unit if required. Let's work on a multiclass classification problem we've seen before. Iris dataset contains elements from three species of Iris flowers. We can create three units in the output layer that will supposedly indicate the probability of an Iris sample falling in one of the three categories.

Let's import the Iris dataset the way we've been doing before.

```
import pandas as pd
import numpy as np
from sklearn.datasets import load_iris
from sklearn.preprocessing import StandardScaler
iris = load_iris()
df = pd.DataFrame(data=iris.data, columns=iris.feature_names)
df['class'] = iris.target
x = df.drop(labels='class', axis=1).astype(np.float32).values
y = df['class'].astype(np.float32).values
```

Because we're going to use PyTorch, let's import the required libraries.

```
import torch, torch.nn as nn
```

We now need to convert x and y to tensors. This can be done using torch.tensor(). Note that we'll also convert the tensors to required data types so that we don't have data format–related issues in the later stages.

```
data = torch.tensor( x ).float()
labels = torch.tensor( y ).long()
print (data.size())
print (labels.size())
Out: torch.Size([150, 4])
     torch.Size([150])
```

We'll now define a simple neural network that accepts four inputs, 16 units in the hidden layer, and three units in the output. All the activations will be ReLU. A schematic diagram for this network is shown in Figure 13-11.

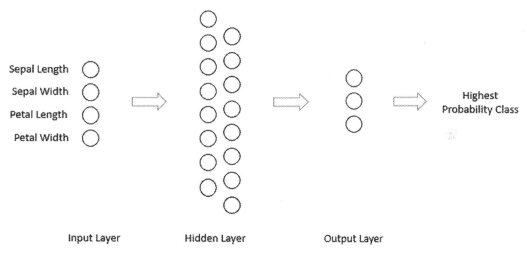

Figure 13-11. *A simple neural network with one hidden layer*

```
model = nn.Sequential(
    nn.Linear(4,16),
    nn.ReLU(),
    nn.Linear(16,16),
    nn.ReLU(),
    nn.Linear(16,3),
      )
```

Let's define loss function and optimizer.

```
crossentropyloss = nn.CrossEntropyLoss()
optimizer = torch.optim.SGD(model.parameters(),lr=.01)
```

Now we can initiate the training loop. In this example, we will train for 1000 iterations, or 1000 epochs. Just like the previous example, we'll keep a track of the losses to visualize the learning process. We'll also compute accuracy by comparing the predictions of the model with the values in the original dataset and keep a record of these for visualizations.

```
maxiter = 1000
losses = []
accuracy = []

for epoch in range(maxiter):
    preds = model(data)
    loss = crossentropyloss(preds,labels)
    losses.append(loss.detach())

    optimizer.zero_grad()
    loss.backward()
    optimizer.step()

    matches = (torch.argmax(preds,axis=1) == labels).
    float()    matchesNumeric = matches.float()
    accuracyPct = 100*torch.mean(matches)
    accuracy.append( accuracyPct )
```

After 1000 iterations, we assume the loss to have sufficiently reduced and accuracies to be consistent. Let's plot the two:

```
import matplotlib.pyplot as plt

fig,ax = plt.subplots(1,2,figsize=(13,4))

ax[0].plot(losses)
ax[0].set_ylabel('Loss')
ax[0].set_xlabel('epoch')
ax[0].set_title('Losses')
```

```
ax[1].plot(accuracy)
ax[1].set_ylabel('accuracy')
ax[1].set_xlabel('epoch')
ax[1].set_title('Accuracy')
plt.show()
```

Figure 13-12 shows the gradual decline of losses that almost didn't vary enough beyond the 800th epoch. The accuracy chart shows a steep increase of accuracy in the initial epochs, which also reached a sufficiently high rate.

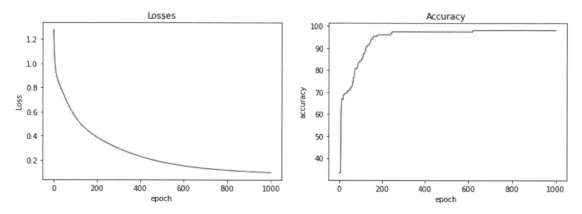

Figure 13-12. *Graphs showing reduction in losses and growth of accuracy over the epochs*

To monitor the accuracy, let's find the predictions again for the final model and compare with the original values:

```
predictions = model(data)
predlabels = torch.argmax(predictions,axis=1)
final_accuracy = 100*torch.mean((predlabels == labels).float())
final_accuracy
Out: tensor(98.)
```

We've achieved sufficiently good accuracy.

However, rather than ending this experiment here, it will be an interesting idea to understand what kind of decision boundaries are created in such a network. This might reveal more insight about how ANNs create boundaries.

We will create a new program for this so that the model is created based on two dimensions of the data (instead of four). Here's the complete code:

```python
import matplotlib.pyplot as plt
import pandas as pd
import numpy as np
from sklearn.datasets import load_iris
from sklearn.preprocessing import StandardScaler
import torch, torch.nn as nn
from matplotlib.colors import ListedColormap

iris = load_iris()
df = pd.DataFrame(data=iris.data, columns=iris.feature_names)
df['class'] = iris.target
x = df.drop(labels='class', axis=1).astype(np.float32).values
y = df['class'].astype(np.float32).values

data = torch.tensor( x[:,1:3] ).float()
labels = torch.tensor( y ).long()

model = nn.Sequential(
    nn.Linear(2,128),    # input layer
    nn.ReLU(),           # activation
    nn.Linear(128, 128), # hidden layer
    nn.Sigmoid(),        # activation
    nn.Linear(128,3),    # output layer
)

crossentropyloss = nn.CrossEntropyLoss()

optimizer = torch.optim.SGD(model.parameters(),lr=.01)

maxiter = 1000

for epochi in range(maxiter):
    preds = model(data)
    loss = crossentropyloss(preds,labels)
    optimizer.zero_grad()
    loss.backward()
    optimizer.step()
```

At this point, the model has been trained, and we can continue preparing a two-dimensional space for plotting a contour plot that will show the decision boundaries based on how the model labels each point.

```
x1_min, x1_max = x[:, 1].min() - 1, x[:, 1].max() + 1
x2_min, x2_max = x[:, 2].min() - 1, x[:, 2].max() + 1
xx1, xx2 = np.meshgrid(np.arange(x1_min, x1_max, 0.01), np.arange(x2_min,
x2_max, 0.01))

predictions = model(torch.tensor(np.array([xx1.ravel(), xx2.ravel()]).
astype(np.float32)).T)
predlabels = torch.argmax(predictions,axis=1)

markers = ('s', 'x', 'o', '^', 'v')
colors = ('red', 'blue', 'lightgreen', 'gray', 'cyan')
cmap = ListedColormap(colors[:len(np.unique(y))])

Z = predlabels.T
Z = Z.reshape(xx1.shape)
plt.contourf(xx1, xx2, Z, alpha=0.3, cmap=cmap)
plt.xlim(xx1.min(), xx1.max())
plt.ylim(xx2.min(), xx2.max())

for idx, cl in enumerate(np.unique(y)):
    plt.scatter(x=x[y == cl, 1],  y=x[y == cl, 2], c=colors[idx],
    marker=markers[idx], alpha=0.5, label=cl,  edgecolor='black')
```

Thus, we'll get a clear plot of how each point in the two-dimensional feature space will be classified as shown in Figure 13-13. On top of that, we have overlaid the original training points.

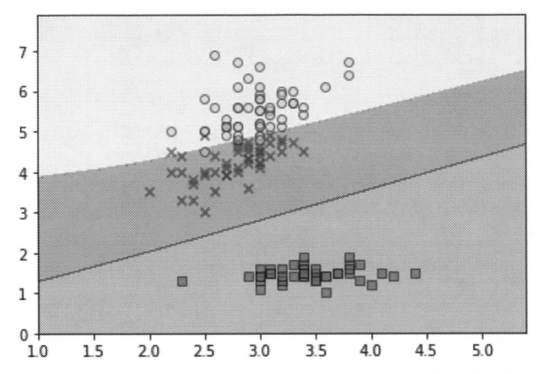

Figure 13-13. *Decision boundaries created by the neural network for classifying Iris data*

Based on the structure of your network and the activation functions, the boundaries might be much more different. However, one interesting pattern here is that the decision boundaries may not be straight – and with more complex data, they might be even more complex.

Now that you understand how neural networks can be built, let's pick another classification problem and work with it.

NN Class in PyTorch

In the previous examples, we have always defined the structure of neural network using nn.Sequential(), which allows us to define how layers and activations are connected to each other. Another way of defining a network is a neural network class that inherits nn.Module and defines the layers that will be used and implements a method to define how forward propagation occurs. This is specifically useful when you want to model a complex model instead of a simple sequence of existing modules.

Let's prepare data for classification using sklearn's make_classification() method to create two distinct clusters.

```
from sklearn.datasets import make_classification
import matplotlib.pyplot as plt

X, y = make_classification(n_samples = 100, n_features=2, n_redundant=0,
n_informative=2, n_clusters_per_class=1, n_classes=2, random_state=15)
plt.scatter(X[:, 0], X[:, 1], marker='o', c=y, s=25, edgecolor='k')
```

We'll now define a simple neural network with an input layer that accepts two inputs: a hidden layer with eight nodes and an output layer with one node denoting the class (0 or 1).

```
import torch, torch.nn as nn, torch.nn.functional as F
import numpy as np

class MyNetwork(nn.Module):
  def __init__(self):
    super().__init__()
    self.input = nn.Linear(2,8)
    self.hidden = nn.Linear(8,8)
    self.output = nn.Linear(8,1)

  def forward(self,x):
    x = self.input(x)
    x = F.relu( x )
    x = self.hidden(x)
    x = F.relu(x)
    x = self.output(x)
    x = torch.sigmoid(x)
    return x
```

In this class, we need to explicitly indicate the sequence of layers and activation functions along with additional operations as we will see in the next section. This class can be instantiated, and model can be trained over multiple epochs in the same way as we did in the previous examples. We add few lines to print a graph to show the change in losses as shown in Figure 13-14.

```
mymodel = MyNetwork()
```

```python
data = torch.tensor(X.astype(np.float32))
labels = torch.tensor(y.reshape(1,100).T.astype(np.float32))
learningRate = .05

lossfun = nn.MSELoss()

optimizer = torch.optim.SGD(mymodel.parameters(),lr=learningRate)

numepochs = 1000
losses = torch.zeros(numepochs)

for epochi in range(numepochs):
  yHat = mymodel(data)
  loss = lossfun(yHat,labels)
  losses[epochi] = loss
  optimizer.zero_grad()
  loss.backward()
  optimizer.step()

# show the losses
plt.plot(losses.detach())
plt.xlabel('Epoch')
plt.ylabel('Loss')
plt.show()
```

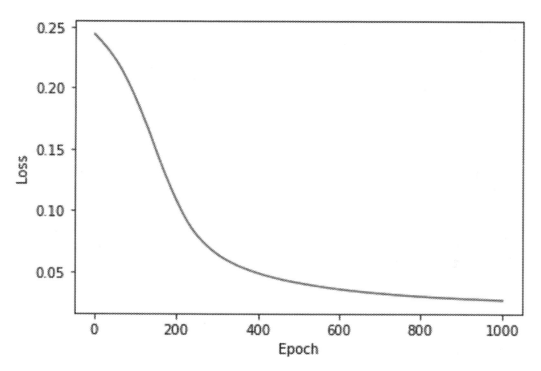

Figure 13-14. *Change in loss for classification over epochs*

Creating such classes helps us define a more sophisticated network block that may be composed of multiple smaller blocks, though in simpler networks, sequential is a good option.

Overfitting and Dropouts

Just like we saw in previous chapters, overfitting is a common problem in machine learning tasks where a model might learn too much from the training dataset and might not generalize well. This is true in neural networks as well, and due to flexible nature, neural networks are susceptible to overfitting.

There are, in general, two solutions to overfitting. One is to use a sufficiently large number of training data examples. The second method requires modifying how complex a network is in terms of network structures and network parameters. You may reduce some layers in the model or reduce the number of computation nodes in each layer.

Another popular method used in neural networks is called dropouts. You can define dropout at a certain layer so that the model will randomly deliberately ignore some of the nodes during training, thus causing those nodes to drop out – and thus, not able to learn "too much" from the training data samples.

Because dropout means less computation, the training process becomes faster, though you might require more training epochs to ensure that the losses are low enough. This method of using dropouts has proven to be effective in reducing overfitting to complex image classification and natural language processing problems.

In PyTorch, dropout can be defined with a probability of dropout, thus

```
dropout = nn.Dropout(p=prob)
```

In the NN class, we can add the dropout layers in the forward propagation definition with a predefined dropout rate of 20%. See the changes in the following code:

```
class MyNetwork(nn.Module):
  def __init__(self):
    super().__init__()
    self.input = nn.Linear(2,8)
    self.hidden = nn.Linear(8,8)
    self.output = nn.Linear(8,1)

  def forward(self,x):
    x = self.input(x)
    x = F.relu( x )
    x = F.dropout(x,p=0.2)
    x = self.hidden(x)
    x = F.relu(x)
    x = F.dropout(x,p=0.2)
    x = self.output(x)
    x = torch.sigmoid(x)
    return x
```

If you keep a track of losses, they might look like Figure 13-15. Though the losses don't reduce smoothly, we eventually find that the loss reduces as we train the network for more and more epochs, and it is possible that the network will be trained with sufficiently usable loss after a certain number of epochs.

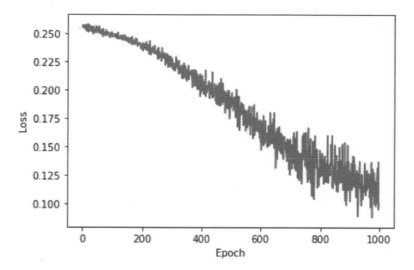

Figure 13-15. *Change in loss function*

Classifying Handwritten Digits

Now that we can create a basic neural network, let's use a more realistic dataset and tackle an image classification problem. For this task, we will use a dataset popularized by Yann LeCun, which contains 70,000 examples of handwritten digits, called MNIST (Modified National Institute of Standards and Technology) database. This dataset has been widely used for image processing and classification. A few digits are shown in Figure 13-16.

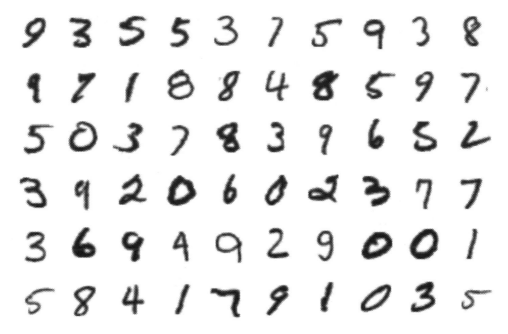

Figure 13-16. *Digits from the MNIST digits dataset*

Let's import the required libraries. For this program, we also need to import torchvision, which is a part of PyTorch and provides various datasets, models, and image transformations.

```
import numpy as np
import matplotlib.pyplot as plt
import torch
import torch.nn as nn
import torch.optim as optim
import torchvision
from torchvision import datasets, transforms
```

Before we download the dataset programmatically, we can define a list of transforms that must be applied to help us process the dataset. We will need to convert the images to tensor format. We can use `torchvision.datasets()` to download the dataset.

```
transform = transforms.Compose([transforms.ToTensor()])

trainset = datasets.MNIST('train', download=True, train=True,
transform=transform)
```

```
testset = datasets.MNIST('test', download=True, train=False,
transform=transform)

train_loader = torch.utils.data.DataLoader(trainset, batch_size=64,
shuffle=True)
test_loader = torch.utils.data.DataLoader(testset, batch_size=64,
shuffle=True)
```

The files will be exported to "/train/MNIST/raw" and "/test/MNIST/raw" folders. The next two lines initialize a DataLoader item for both training data and test data. DataLoader doesn't directly provide the data but can be controlled by a user-defined iterable. Before we define a neural network, let's look at a data item. The following lines would generate a batch of training data of 64 elements.

```
dataiter = iter(train_loader)
images, labels = dataiter.next()
```

You can locate a picture using

```
plt.imshow(images[0].numpy().squeeze(), cmap='gray_r');
```

Each digit is present in a 28x28 pixel box. Each pixel, in this grayscale dataset, contains a value from 0 to 255 indicating the color (in terms of darkness). Thus, each image is defined by 784 values. Now we can proceed to defining the network. We know that the input layer contains 784 units and the output layer contains 10 units, each representing a number. We will add two more input layers with 64 units each. For this example, we will keep activations as ReLU, and because we have a multiclass classification problem, loss layer will be cross-entropy. Figure 13-17 summarizes the neural network we will be creating in this example.

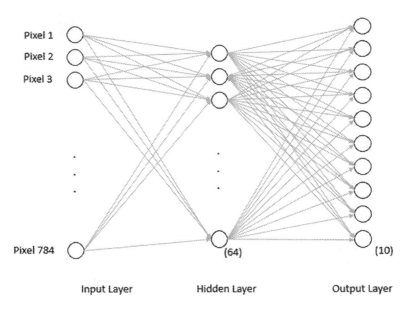

Figure 13-17. *Neural network for classifying handwritten digits. The input will always be of length 784, and the output will have a length of 10, each representing a digit*

```
model = nn.Sequential(nn.Linear(784, 64),
                      nn.ReLU(),
                      nn.Linear(64,64),
                      nn.ReLU(),
                      nn.Linear(64,10),
)
print(model)

Out: Sequential(
  (0): Linear(in_features=784, out_features=64, bias=True)
  (1): ReLU()
  (2): Linear(in_features=64, out_features=64, bias=True)
  (3): ReLU()
  (4): Linear(in_features=64, out_features=10, bias=True)
)
```

We now define the loss function and the optimizer.

```
lossfn = nn.CrossEntropyLoss()
optimizer = torch.optim.SGD(model.parameters(),lr=.01)
```

In the training phase, we will limit the number of epochs to ten, as the dataset is much larger than the examples we've covered so far. However, this should be good enough to bring the losses sufficiently low to make the model able to predict well for most of the examples. Within each epoch, train_loader would iterate over batches of 64 entries.

```
losses = []
for epoch in range(10):
    running_loss = 0
    for images, labels in train_loader:
        images = images.view(images.shape[0], -1)
        optimizer.zero_grad()
        output = model(images)
        loss = lossfn(output, labels)
        loss.backward()
        optimizer.step()
        running_loss += loss.item()
    print("Epoch {} - Training loss: {}".format(epoch, running_loss/
    len(train_loader)))
    losses.append(running_loss/len(train_loader))
```

This should show gradual reduction of losses over each epoch.

```
Epoch 0 - Training loss: 1.7003767946635737
Epoch 1 - Training loss: 0.5622020193190971
Epoch 2 - Training loss: 0.4039541946005211
Epoch 3 - Training loss: 0.35494661225534196
Epoch 4 - Training loss: 0.32477016467402486
Epoch 5 - Training loss: 0.302617403871215
Epoch 6 - Training loss: 0.2849765776086654
Epoch 7 - Training loss: 0.2697247594261347
Epoch 8 - Training loss: 0.25579357369622185
Epoch 9 - Training loss: 0.24312975907773732
```

The graph in Figure 13-18 shows how losses reduce over time.

```
plt.plot(losses)
plt.xlabel('Epoch')
plt.ylabel('Loss')
plt.show()
```

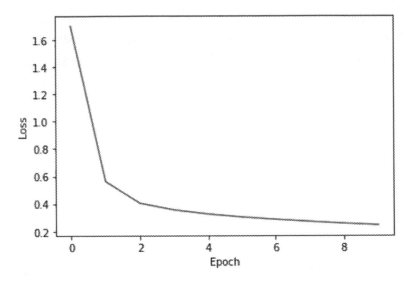

Figure 13-18. *Change in loss over a few epochs*

Let's now test how the model predicts for an item from the test dataset:

```
testimgs, testlabels = iter(test_loader).next()
plt.imshow(testimgs[0][0].numpy().squeeze(), cmap='gray_r');
```

Figure 13-19 shows the first digit from the test dataset. The sample we had clearly shows that this is the number eight (8).

Please note that some samples in MNIST datasets might not be so clear. As you might expect, the dataset contains some examples that might be confusing to even human readers, for example, a six (6) that looks more like a zero (0) or a seven (7) that looks more like a one (1). Some simple neural networks trained for a small number of epochs might not be very accurate.

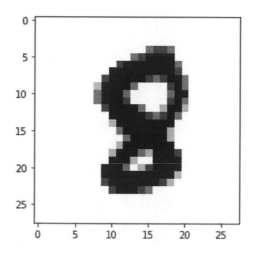

Figure 13-19. *A single digit from MNIST dataset*

We will first compress the image to 1x784 cell format and find the values at the output layer. The highest value would correspond to the predicted value. We'll also convert the values of output cells to the probabilities, of which, the best can be chosen as the output.

```
img = testimgs[0].view(1, 784)
with torch.no_grad():
    logps = model(img)

ps = torch.exp(logps)
probabilities = list(ps.numpy()[0])
prediction = probabilities.index(max(probabilities))
print(prediction)
Out: 8
```

Here, all the ten output units represent a probability. You can explore probabilities object to see the probability of alternate outputs – we can see that this eight (8) also has a very minor chance of being a 6 or a 2.

Summary

In this chapter, we learned how neural network units work and how they can be combined to create more capable neural networks that can solve complex problems. We created simple and complex neural networks for regression and classification problems. In the next chapter, we'll proceed to a special kind of neural network architecture especially suitable for images and other two- or three-dimensional data.

## CHAPTER 14

Convolutional Neural Networks

A convolutional neural network (CNN) is a feedforward neural network with layers for specialized functions for applying filter to the input image by sliding a filter across small sections of the image to produce an activation map. Recall that regular feedforward networks are made up of individual computation units or neurons, and the training process requires backpropagation to learn their weights and biases. Each computation unit combines the input with weights using a dot product and passes it to a nonlinear function that produces an output. This output is often used as input to the neurons in the next layer till we reach the output layer. CNNs are developed and centered around finding features within images with the use of a mathematical operation called convolution in at least one of their layers.

In this chapter, we will understand how CNNs work, followed by discussion on individual parts of CNN that may be tuned, and see examples of convolutional neural networks in action.

Convolution Operation

Each pixel in an image is not a unique data source but is dependent on the pixels around it – often portraying continuity of the same object by having similar values (similar color or intensity), or abrupt boundary of object by sharp difference in the values in terms of color or brightness. Convolution allows us to capture such features through a set of predefined kernels that operate on the images.

Convolution is a mathematical operation in which a kernel (or filter) is applied to a part of the input data (usually image) to produce an element, slid over the whole image, to produce a final modified image. Consider the matrix and the filter as shown in Figure 14-1.

© Ashwin Pajankar and Aditya Joshi 2022
A. Pajankar and A. Joshi, *Hands-on Machine Learning with Python*, https://doi.org/10.1007/978-1-4842-7921-2_14

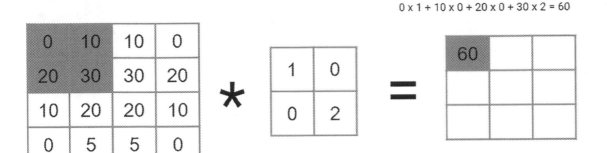

Figure 14-1. *Part of convolution operation with 4 x 4 matrix and 2 x 2 filter*

The 4x4 matrix is operated with a 2x2 filter, which leads to the filter being mapped to the top-left part of the matrix. This leads to one singular value that is saved as the output in one of the cells of the resulting matrix as shown in Figure 14-1. The filter then slides to the right and performs a similar computation to the next part of the image shown in Figure 14-2.

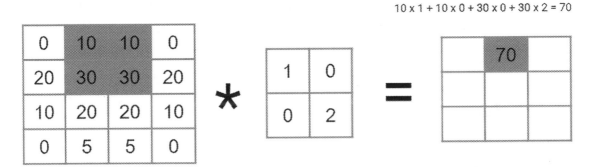

Figure 14-2. *Subsequent part of convolution operation with 4 x 4 matrix and 2 x 2 filter*

The process continues till all the values of the original matrix are filled as shown in Figure 14-3.

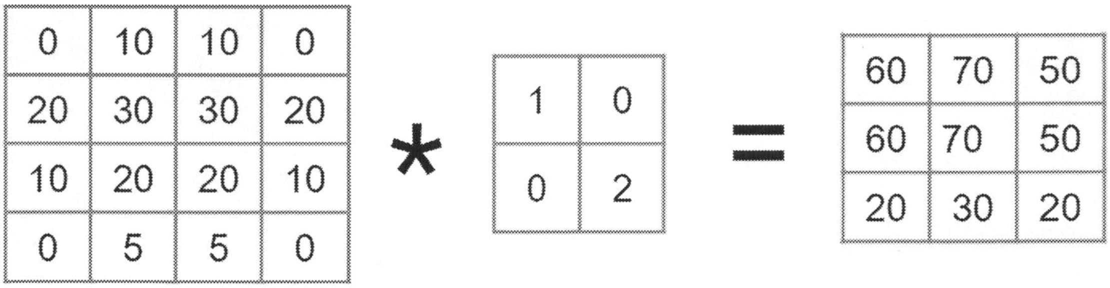

0	10	10	0
20	30	30	20
10	20	20	10
0	5	5	0

*

1	0
0	2

=

60	70	50
60	70	50
20	30	20

Figure 14-3. *Result of convolution operation with 4 x 4 matrix and 2 x 2 filter*

As the resulting matrix is much smaller, we sometimes add padding so that the resulting image is of similar size as that of the original image. So we may instead begin with the padded matrix as shown in Figure 14-4.

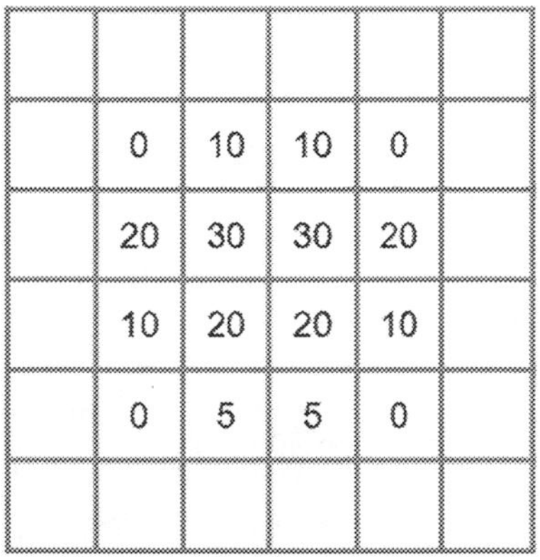

	0	10	10	0	
	20	30	30	20	
	10	20	20	10	
	0	5	5	0	

Figure 14-4. *Padding to ensure that the convolution results are of the same shape*

Filter essentially filters out everything that is not related to the pattern contained in the filter. We want to learn the filters to find patterns based on the given dataset. In easy words, convolution is just a sliding pattern finder. In image processing software, there are some standard filters that provide the common effects like sharpening, blurring, edge detection, etc. A simple example is shown in Figure 14-5.

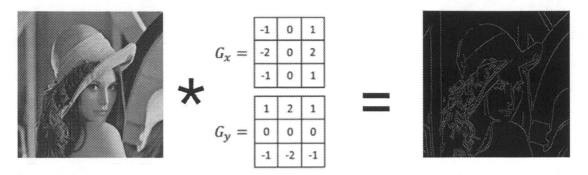

Figure 14-5. *Filter for detecting parts of the image*

We can build more sophisticated filters that can be used to discover a particular type of patterns. Figure 14-6 shows a filter containing a pattern (or image) of a human eye, and when a convolution operation is performed on an image with this filter, the result will contain a high value where the filter closely matches a pattern in the original image.

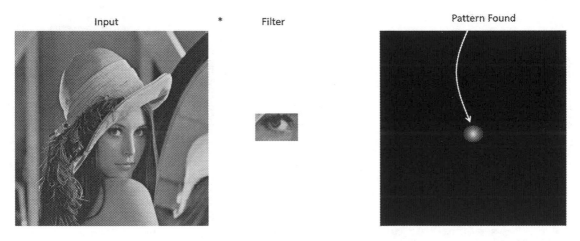

Figure 14-6. *Filter for detecting patterns in image*

In colored images, each image is a 3xrxh three-dimensional array where r is the number of rows and h is the height of the image in terms of pixels. There are three channels of rxh size matrix that represent the red, green, and blue channels of the image. For such a case, the kernel also needs to be a three-channel kernel. Thus, if we had a kernel of size 3x3 in two dimensions, now we would have a kernel of size 3x3x3, where we apply layers of kernel in each channel of the image.

On applying the filter to the image, the resulting matrix will be a two-dimensional image as shown in Figure 14-7.

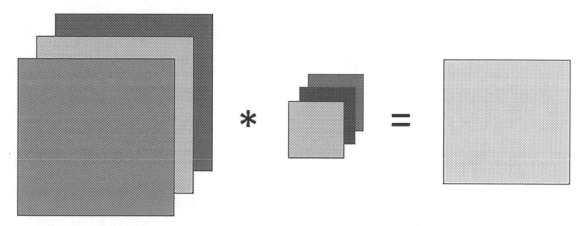

Figure 14-7. *Filters across multiple channels produce a single-channel output*

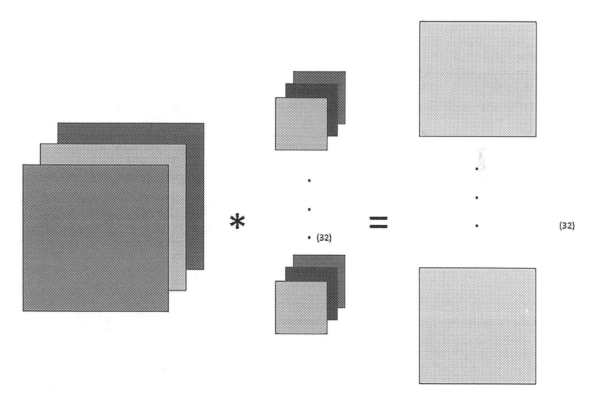

Figure 14-8. *Multiple filters lead to multiple outputs*

We can have more than one kernel. Each kernel will produce one two-dimensional matrix as shown in Figure 14-8. Thus, if we use three different kernels, the result will be a three-dimensional matrix that has result of each kernel present as layers across one of the dimensions.

It is important to note that many libraries implement a function called cross-correlation instead. As we saw in the previous chapter, the parameters are learned by routing back the gradients during backpropagation. In CNNs, the backward pass used the same but flipped convolution operation with the same parameter to learn the right weights of the filters.

Structure of a CNN

CNNs are neural networks that leverage convolution operation in at least one of the layers. There are many popular architectures that have been proven to be effective in various image classification–related tasks. Figure 14-9 shows a simple and the most standard configuration of convolutional neural networks.

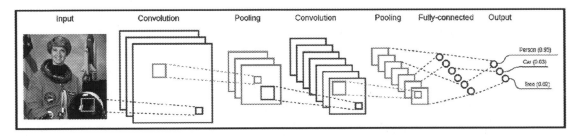

Figure 14-9. *A standard configuration of convolutional neural networks*

The input is a H x W x C shape tensor that holds the original values of image pixels. H and W represent the height and weight of the image in terms of pixels, and C represents the number of channels (usually three for RGB color image and one for monochromatic grayscale images). The input is connected to the convolution layer that performs convolution operation on the image with F filters. Thus, with a padding of 1 and a stride of 1 (explained in the next paragraphs), this will lead to a H x W x F size output. This is followed by an activation layer (preferably ReLU) that doesn't affect the size.

After convolution and activation, a pooling layer can compress the feature activation information about the image by aggregating nearby pixels and down sample the data. Max pooling can be used to pick the maximum value of each part of the image. It helps make the representation translation invariant, that is, not being affected by small changes in location of a feature in the image. Thus, H x W x F image may result in H/2 x W/2 x F sized output.

At this point, we may decide to flatten this and forward to one or more arbitrary fully connected layers with C units in the output layer, where C is the number of categories the image input may be classified into.

In many applications, we add multiple sets of convolution layer followed by polling layer, which can aggregate multiple features that are detected in the previous stage.

Padding and Stride

When specifying the hyperparameters of filters in a convolution layer, we may choose a depth, stride, and padding. These decisions may have major implication in the performance and capabilities of the final model thus being generated.

Depth, or the number of filters, decides how many different features within the image may be captured by the network. Each filter may learn to discover a pattern in the image, a certain kind of edges, or color combinations. While specifying this, we also specify the size of the filters.

Stride defines how much the filter slides across the original image. A default stride of 1 means that the filter moves only one pixel at a time. A stride of 2 means that the filter will jump over by two steps instead and then perform the multiplication and summation operations at the next location in the image.

Figure 14-10 shows the difference in the output because of change in strides.

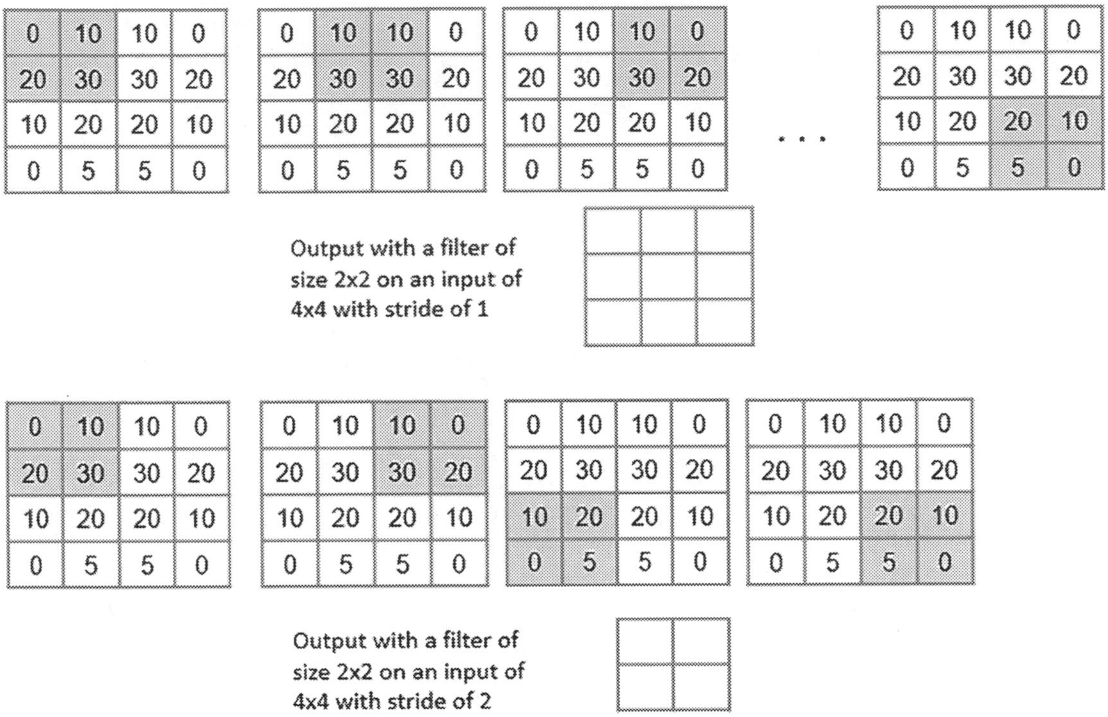

Output with a filter of
size 2x2 on an input of
4x4 with stride of 1

Output with a filter of
size 2x2 on an input of
4x4 with stride of 2

Figure 14-10. *Difference in the shape of the output because of different stride*

Padding adds another rows and columns at the borders of the original image and puts a value of 0 in these cells. As a convolution operation combines the information from a larger part of the image, a padding of 1 with a stride of 1 would result in the output of the same size as that of the input image.

CNN in PyTorch

PyTorch provides multiple classes for implementation of convolution layer. Convolution can happen by sliding a filter across one direction or multiple directions on the image. 1D convolution has been found to be highly effective in conjunction with other neural network layers on text classification tasks in which text embeddings can be used as a representation of the original text. It can be used with images as well in case you are particularly interested in capturing features across an axis; however, in most cases, 2D convolution is more suitable.

In PyTorch's implementation of 2D convolution, you have to define input channels, output channels, and kernel size. Default stride is 1 and padding is 0. Let's import the required modules and define a simple 2-DConv layer.

```
from torch import nn
conv = nn.Conv2d(3,15,3,1,1)
conv
Out: Conv2d(3, 15, kernel_size=(3, 3), stride=(1, 1), padding=(1, 1))
```

This creates a conv layer that accepts three channels as input, 15 channels in the output for a kernel of shape 3x3. The kernel will operate on the original image with additional rows and columns and slide with one-pixel step at a time.

There are 15 filters of size 3x3 each across each of the three channels. To check the shape of weights to be learned, you can use the following:

```
conv.weight.shape
Out: torch.Size([15, 3, 3, 3])
```

To understand how filters operate, let's run a simple filter convolution. Let's import an image[1] and use a simple filter and see how it can affect the image.

```
import urllib.request
from PIL import Image

urllib.request.urlretrieve( 'https://upload.wikimedia.org/wikipedia/
commons/thumb/2/26/Boat_in_the_beach_Chacachacare.jpg/640px-Boat_in_the_
beach_Chacachacare.jpg', "boat.png")
img = Image.open("boat.png")
```

The preceding lines download the jpg and save the image locally as a PNG file. We then open the PNG. You can load the img as a NumPy array to see the values at individual pixels using np.array(img). Note that the structure of the image we loaded puts the channel information at the innermost dimension – which is different when we load the image in Torch.

```
from PIL import Image
import torchvision.transforms.functional as TF
```

---

[1]https://upload.wikimedia.org/wikipedia/commons/thumb/2/26/Boat_in_the_beach_
Chacachacare.jpg/640px-Boat_in_the_beach_Chacachacare.jpg

```
x = TF.to_tensor(img)
x.unsqueeze_(0)
print(x.shape)
```

```
Out: torch.Size([1, 3, 378, 640])
```

PyTorch thus confirms that we have loaded one image with three channels containing information about an image that is 378 pixels high and 640 pixels wide.

Let's define a simple filter that will boost the contrast sharply:

```
imgfilter = torch.zeros((1,3,3, 3))
imgfilter[0,:]=torch.tensor([[-10,10,-10],[10,100,10],[-10,10,-10]])
imgfilter = torch.nn.Parameter(imgfilter)
```

Let's see the effect of convolution:

```
import torch.nn.functional as F
z = F.conv2d(x, imgfilter, padding=1, stride=1)
z = z.detach().numpy()[0][0]
output = Image.fromarray(z)
output.show()
```

In this implementation, z is originally created as a 1x1x378x640 array, of which we extract the actual image. This produces a sharp highlighted grayscale image as shown in Figure 14-11.

Figure 14-11. *Sharp grayscale image produced by a custom filter*

In this process, we want to learn filters that help us find the specific features of the image that may lead to an improved (reduced) loss.

Let's bring it all together and create a neural network using convolution layers to classify images.

Image Classification Using CNN

In this example, we will use another popular dataset called MNIST Fashion. MNIST Fashion attracted a lot of attention due to debate against the use of MNIST digits as many relatively simple CNN-based networks were easily able to achieve accuracies of 99% or more. Ian Goodfellow, a highly cited deep learning researcher working with Apple Inc. (as of mid-2021), made a comment on social media saying "Instead of moving on to harder datasets than MNIST, the ML community is studying it more than ever." This led to exploration of other datasets, and Zalando ten-category clothing classification dataset soon caught public attention.

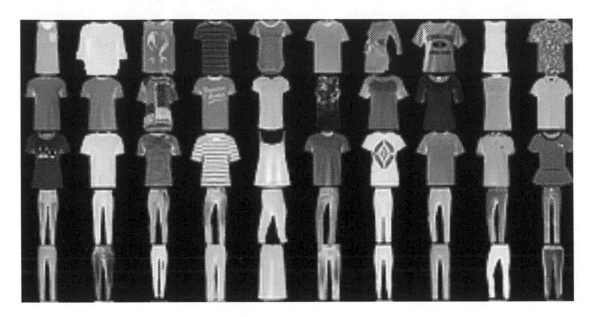

Figure 14-12. *Items from the MNIST Fashion dataset*

Each image in the dataset is a 28x28 pixel grayscale image, and each pixel is an integer between 0 and 255. There are 60000 items in the training set and 10000 items in the test set, each belonging to one of these categories: T-shirt/top, Trouser, Pullover, Dress, Coat, Sandal, Shirt, Sneaker, Bag, and Ankle boot. A small part of the dataset is shown in Figure 14-12.

In this example, we will build a CNN-based neural network with ten units at the output layer that will help us classify a 28x28 pixel grayscale image as one of the ten categories. Torchvision datasets include API to download MNIST Fashion dataset. Let's begin by importing the required modules.

```
import numpy as np
import pandas as pd
import matplotlib.pyplot as plt

import torch
import torch.nn as nn
from torch.autograd import Variable

import torchvision
import torchvision.transforms as transforms
from torch.utils.data import Dataset, DataLoader
from sklearn.metrics import confusion_matrix
```

We'll need to define transforms to convert the data to the same required shape and format. We'll then call FashionMNIST() to initialize the training and testing datasets which will download the data to the mentioned folder fashion_data if the data is not present.

```
all_transforms = transforms.Compose([
        transforms.ToTensor()
    ])
train_data = torchvision.datasets.FashionMNIST ('fashion_data', train=True,
download=True, transform=all_transforms)
test_data = torchvision.datasets.FashionMNIST ('fashion_data', train=False,
transform=all_transforms)
```

Out: Downloading http://fashion-mnist.s3-website.eu-central-1.amazonaws.com/train-images-idx3-ubyte.gz

Downloading http://fashion-mnist.s3-website.eu-central-1.amazonaws.com/train-images-idx3-ubyte.gz to fashion_data_train\FashionMNIST\raw\train-images-idx3-ubyte.gz

...

This will install around 80MB compressed data. We can now initialize DataLoader() objects with a predefined batch_size.

```
train_loader = DataLoader(train_data, batch_size=64, shuffle=True)
test_loader = DataLoader(test_data, batch_size=64, shuffle=True)
```

We would recommend you to modify the batch_size as higher batch_sizes might also be appropriate in the systems with sufficient resources. Let's check one batch of data items.

```
samples, labels = next(iter(train_loader))
samples.size()
Out: torch.Size([64, 1, 28, 28])
```

This confirms that samples contain 64 elements, containing each sample of 28x28 shape. There's only one grayscale channel in each image. Labels contain a tensor of size [64] that contains a number from 0 to 9 to represent the class each item falls in. Let's create a mapping of MNIST Fashion labels so that we can refer to it when required:

```
def mnist_label(label):
    output_mapping = {
                0: "T-shirt/Top",
                1: "Trouser",
                2: "Pullover",
                3: "Dress",
                4: "Coat",
                5: "Sandal",
                6: "Shirt",
                7: "Sneaker",
                8: "Bag",
                9: "Ankle Boot"
                }
    label = (label.item() if type(label) == torch.Tensor else label)
    return output_mapping[label]
```

Let's try one item from the samples and check its label.

```
idx = 2
plt.imshow(samples[idx].squeeze(), cmap="gray")
print (mnist_label(labels[idx].item()))
```

You may check the other elements in the dataset by changing the value of the preceding idx variable. The output for an item is shown in Figure 14-13.

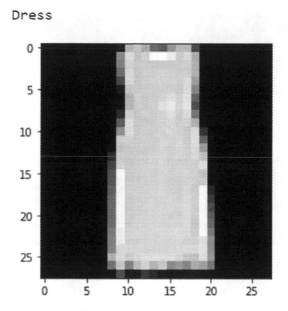

Figure 14-13. One of the items from MNIST Fashion dataset

To explore the data further, you can visualize multiple items in a grid. The following segment of code will produce an output similar to Figure 14-14.

```
sample_loader = torch.utils.data.DataLoader(train_data, batch_size=10)

batch = next(iter(sample_loader))
images, labels = batch

grid = torchvision.utils.make_grid(images, nrow=10)

plt.figure(figsize=(15, 20))
plt.imshow(np.transpose(grid, (1, 2, 0)))

for i, label in enumerate(labels):
    plt.text(32*i, 45, mnist_label(label))
```

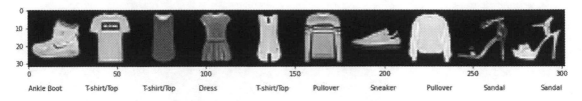

Figure 14-14. *MNIST items loaded in the training dataset with their labels*

Now that we have a clearer understanding of the data, we can proceed to design a neural network. In such networks, an interesting aspect to keep in mind is the sizes of tensors we'll have as input and output in each layer.

The input layer consists of a 1x28x28 pixel image. This will connect with a convolutional layer (conv2d), which takes input with one channel and produces 32 channels through 32 filters. This number is configurable and determines the complexity of this layer. Within this layer, we can use ReLU as activations. To simplify and make sure that the features extracted are more generic and translation invariant, we will add a two-dimensional max pooling layer. The size of max pooling will affect the design of following layers. Figure 14-15 summarizes the structure of the convolution part of the network.

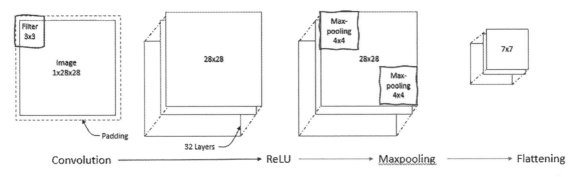

Figure 14-15. *Schematic diagram showing operations in a convolutional neural network*

This can be defined as

```
convlayer = nn.Sequential(
    nn.Conv2d(in_channels=1, out_channels=32, kernel_size=3,
    padding=1),
    nn.ReLU(),
    nn.MaxPool2d(kernel_size=4, stride=4)
)
```

Max pooling leads to generation of a 32x7x7 tensor, which we can forward to a dense network. We will create a dense layer with 32x7x7 inputs and arbitrarily 64 outputs. This also defines the next layer to be limited to 64 units. We know that output layer needs to have ten outputs. Figure 14-16 simplifies the final network we thus create.

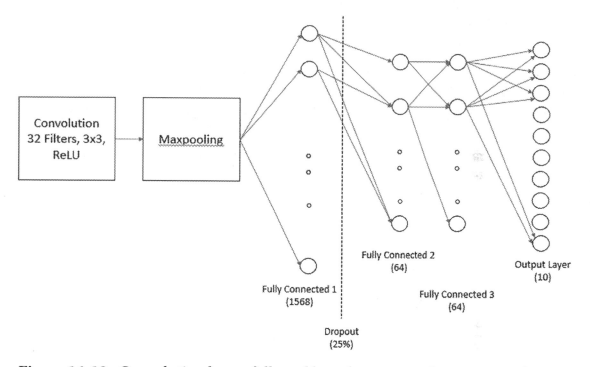

Figure 14-16. *Convolution layers followed by a dense network consisting of multiple fully connected layers*

Let's create the model class for the network. As we saw in the previous chapter, we need to define the individual layers in the constructor and connect them together in a method called forward(), which will compute the outputs in the forward propagation phase. We will use the same definition of convolution layer as we saw in the previous page. Here's the consolidated class.

```
class FashionCNN(nn.Module):

    def __init__(self):
        super(FashionCNN, self).__init__()

        self.convlayer = nn.Sequential(
```

```
                nn.Conv2d(in_channels=1, out_channels=32, kernel_size=3,
                padding=1),
                nn.ReLU(),
                nn.MaxPool2d(kernel_size=4, stride=4)
        )

        self.fully_connected_layer_1 = nn.Linear(in_features=32*7*7, out_
        features=64)
        self.drop = nn.Dropout2d(0.25)
        self.fully_connected_layer_2 = nn.Linear(in_features=64, out_
        features=64)
        self.fully_connected_layer_3 = nn.Linear(in_features=64, out_
        features=10)

    def forward(self, x):
        out = self.convlayer(x)
        out = out.view(out.size(0), -1)
        out = self.fully_connected_layer_1(out)
        out = self.drop(out)
        out = self.fully_connected_layer_2(out)
        out = self.fully_connected_layer_3(out)
        return out
```

You can see the details of the model thus being created by initializing an object of the class.

```
model = FashionCNN()
print(model)

Out:
FashionCNN(
  (convlayer): Sequential(
    (0): Conv2d(1, 32, kernel_size=(3, 3), stride=(1, 1), padding=(1, 1))
    (1): ReLU()
    (2): MaxPool2d(kernel_size=4, stride=4, padding=0, dilation=1, ceil_
    mode=False)
  )
```

```
(fully_connected_layer_1): Linear(in_features=1568, out_features=64,
bias=True)
(drop): Dropout2d(p=0.25, inplace=False)
(fully_connected_layer_2): Linear(in_features=64, out_features=64,
bias=True)
(fully_connected_layer_3): Linear(in_features=64, out_features=10,
bias=True)
)
```

We can now initialize the loss function and optimizer and create a general loop for the training process. We will keep a track of the loss and accuracies as we've done before.

```
error = nn.CrossEntropyLoss()
optimizer = torch.optim.Adam(model.parameters(), lr=0.005)
lstlosses = []
lstiterations = []
lstaccuracy = []
```

However, we can also use test data to find the test accuracy and track how train and test accuracy changed over the training period. The training loop will take care of three items: (1) iterate over train_loader and run forward and backpropagation to optimize the parameters, (2) iterate over test loader to predict labels of test data items and keep a log of test accuracy, and because the training process will be slow, (3) print or log a statement indicating the progress of the training process. Let's create additional objects to track test items.

```
predictions_list = []
labels_list = []

num_epochs = 10 # Indicate maximum epochs for training
num_batches = 0 # Keep a track of batches of training data
batch_size = 100 # Configurable for tracking accuracy

for epoch in range(num_epochs):
    print ("Epoch: {} of {}".format(epoch+1, num_epochs))
    for images, labels in train_loader:
        train = Variable(images)
        labels = Variable(labels)
```

```python
        outputs = model(train)
        loss = error(outputs, labels)
        optimizer.zero_grad()
        loss.backward()
        optimizer.step()
        num_batches += 1

        if num_batches % batch_size==0:
            total = 0
            matches = 0

            for images, labels in test_loader:
                labels_list.append(labels)
                test = Variable(images)
                outputs = model(test)

                predictions = torch.max(outputs, 1)[1]
                predictions_list.append(predictions)
                matches += (predictions == labels).sum()
                total += len(labels)

            accuracy = matches * 100 / total
            lstlosses.append(loss.data)
            lstiterations.append(num_batches)
            lstaccuracy.append(accuracy)

        if not (num_batches % batch_size):
            print("Iteration: {}, Loss: {}, Accuracy: {}%".format(num_
            batches, loss.data, accuracy))
```

This block should take time and print the loss and accuracy information intermittently.

```
Out:
Epoch: 1 of 10
Iteration: 50, Loss: 0.6086341142654419, Accuracy: 76.94999694824219%
Iteration: 100, Loss: 0.5399684906005859, Accuracy: 79.62000274658203%
Iteration: 150, Loss: 0.46176013350486755, Accuracy: 82.83000183105469%
Iteration: 200, Loss: 0.3623868227005005, Accuracy: 83.23999786376953%
```

```
Iteration: 250, Loss: 0.26515936851501465, Accuracy: 85.12999725341797%
...
Iteration: 900, Loss: 0.4560268521308899, Accuracy: 86.41000366210938%
Epoch: 2 of 10
Iteration: 950, Loss: 0.3548094928264618, Accuracy: 86.93000030517578%
Iteration: 1000, Loss: 0.18031403422355652, Accuracy: 87.3499984741211%
Iteration: 1050, Loss: 0.35414841771125793, Accuracy: 87.13999938964844%
Iteration: 1100, Loss: 0.2871503233909607, Accuracy: 87.5199966430664%
...
```

For further experimentation and improvement of accuracy, you may choose to increase the number of conv2d layers or fully connected layers. Another factor to consider is the size of filters, max pooling length, and number of units in the fully connected layers. You can add more convolution layers and configure your network as follows:

```
self.convlayer_2 = nn.Sequential(
    nn.Conv2d(in_channels=1, out_channels=32, kernel_size=3,
    padding=1),
    nn.ReLU(),
    nn.MaxPool2d(kernel_size=4, stride=4)
)
```

Changes like this often require you to reconsider the shape of every tensor that might be affected. Due to simplicity provided by PyTorch API, you only need to add how convlayer_2 will fit in forward(), and loss.backward() and optimizer.step() will update the parameters accordingly.

As we have logged losses and accuracy during training, we can understand how the model performed over the epochs by visualizing them using Matplotlib.

```
plt.plot(lstiterations, lstlosses)
plt.xlabel("No. of Iteration")
plt.ylabel("Loss")
plt.title("Iterations vs Loss")
plt.show()

plt.plot(lstiterations, lstaccuracy)
plt.xlabel("Iterations")
```

```
plt.ylabel("Accuracy")
plt.show()
```

The plots in Figure 14-17 indicate that though there is further scope of improvement, the losses have reduced over epochs and the accuracy has increased.

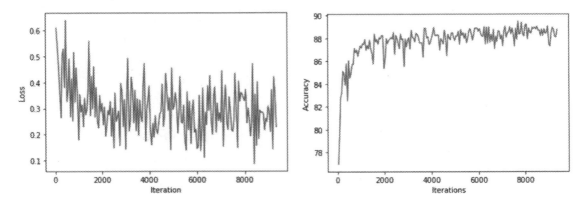

Figure 14-17. *Changes in loss and accuracy across multiple iterations*

What Did the Model Learn?

At any point, a model's parameters (weights and biases) can be accessed using `model.parameters()`, which returns a generator for tensors that can be iterated over to get all the weights. The model contains parameter values as soon as it is initialized – mostly as random values, which get updated during the backpropagation.

```
list(model.parameters())[0].shape
Out: torch.Size([32, 1, 3, 3])
```

The tensor returned at index 0 contains the 32 filters for convolution layer, present in the form of single-channel (grayscale) 3x3 size image. Let's read their values and visualize to see what the convolution layer actually learned.

```
filters = list(model.parameters())[0]
filters

Out: Parameter containing:
tensor([[[[ 2.7956e-01, -2.4780e-01, -2.0801e-01],
          [ 2.3427e-01, -1.0169e-01, -2.2255e-01],
          [-8.3205e-02,  4.7983e-02,  2.8062e-01]]],
```

```
[[[ 1.4788e-02, -3.9390e-02, -9.7228e-02],
  [-2.4128e-01, -2.5705e-01,  2.2378e-01],
  [ 1.3481e-01, -2.4824e-01, -2.8518e-01]]],
```
...

This is truncated output showing the first two filters. We can visualize these values using Matplotlib. The output is shown in Figure 14-18.

```
plt.figure(figsize=(20, 17))
for i, filter in enumerate(filters):
    plt.subplot(8, 8, i+1)
    plt.imshow(filter[0, :, :].detach(), cmap='gray')
    plt.axis('off')
plt.show()
```

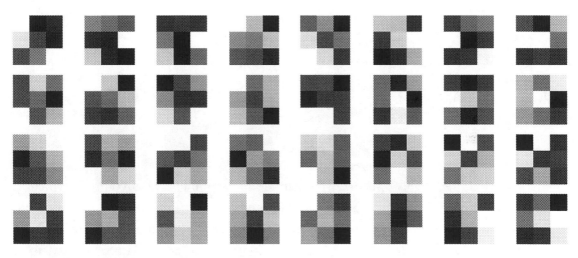

Figure 14-18. *Filters visualized*

Filters of larger size or filters of subsequent convolution layers can even show more understandable features like collar of a shirt or heel of footwear. In many practical applications, more sophisticated networks with multiple convolution layers are built, which are followed by dense layers at the later end.

Deep Networks of CNN

CNNs were originally devised in the mid-1990s but found limited applications due to the state of the art of computing infrastructure of that time. In the last decade, CNNs were highlighted again due to high performance in complex classification datasets through deep networks involving multiple convolution layers.

Of them, Alex Krizhevsky's network, called AlexNet, is considered as one of the most popular and influential papers that gave the future direction to computer vision. This network is used to classify ImageNet, which is a large database of images categorized in thousands of categories. Figure 14-19 shows the structure of AlexNet, which was published in the original paper.

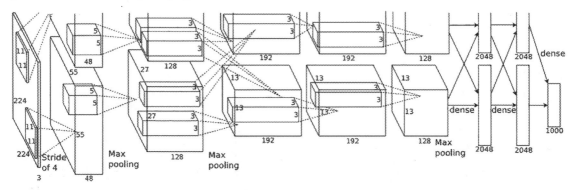

Figure 14-19. *Schematic diagram of AlexNet*

AlexNet included highly optimized GPU implementation of 2D convolution, which was used in five convolutional layers, followed by max pooling and a dense network. The output layer was a 100-unit layer that produces a distribution over the 1000 class labels. They leveraged dropout layers that had been recently popularized by Hinton et al. with a probability of 0.5 to overcome overfitting. Such dropouts were believed to force neurons to not rely on the presence of particular other neuron and learn more robust features.

Summary

In this chapter, we've learned a highly specialized type of neural network architectures called convolutional neural network (CNN) and implemented them for image classification examples. In the following chapter, we will study another architecture that is highly effective for sequential data like text and speech due to implementation of the memory that holds a state. These recurrent neural networks (RNN) have different types based on the use cases they are applied on.

CHAPTER 15

Recurrent Neural Networks

In the previous chapters, we've seen that simple data that can be expressed as a vector can be easily provided to the input layers of a feedforward neural network. More complex data like images can either be transformed and flattened to be sent as input in vector form or can be used to learn filters in a convolutional neural network (CNN). CNNs helped capture essential patterns in the data that occur due to certain values present in the proximity of a pixel. However, there's another pattern that usually occurs in data formats like text, speech, etc., as shown in Figure 15-1.

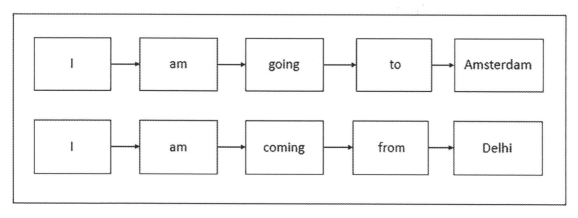

Figure 15-1. *Text as a sequence of tokens*

In sequential data formats like these, it can be observed that in terms of steps of time (say, based on time slice in audio, or character by character in text), the patterns between the data at the current step and the previous step(s) lead to a different meaning if the data in one of the steps changes.

© Ashwin Pajankar and Aditya Joshi 2022
A. Pajankar and A. Joshi, *Hands-on Machine Learning with Python*, https://doi.org/10.1007/978-1-4842-7921-2_15

Recurrent neural networks (RNNs) are a type of neural networks that are specialized for processing a temporal sequence of data. In this chapter, we will learn about the structure of the basic unit of the RNNs and how they fit in a larger neural network. We'll also study specialized RNN units like LSTM that have broken records for machine translation, text-to-speech generation, speech recognition, and other tasks.

Recurrent Unit

Recurrent neural networks accept the provided chunk of input data and use an internal state to produce the output. The internal state is calculated from the old internal state and the input – thus, we can say that indirectly, an output of the recurrent cell in the neural network in a previous step determines the output in the current step. In a way, if we understand it in this way, recurrent neural networks get part of their output as input for the next time step.

Assume that the data obtained at a particular timestamp is present in the form of D-dimensional vectors. At time step index 1, x(1) is the input vector of size D, followed by x(2) in time step index 2, and so on, up to x(T) at time step T. This represents one sample of the dataset in the form of a TxD shape matrix as shown in Figure 15-2. For N samples like these, you can assume the data in the form of an NxTxD shape matrix as shown in Figure 15-2.

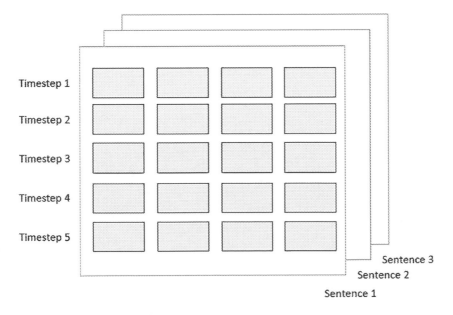

Figure 15-2. *Text represented as vectors across time steps*

At the simplest level, input at each unit is applied to a weighted function as you saw in the previous chapters. These weights are shared through time and applied to each input, along with the output of the previous step. This operation can be expressed as shown in Figure 15-3.

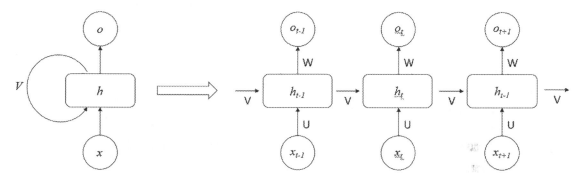

Figure 15-3. *Unfolding of an RNN cell across time steps*

The portion on the left shows that the input x passed to the recurrent unit cell applied with the weights W leads to the output o. When this recursion is unfolded over time as shown in the portion on the right, you can see that there are three time steps, each with inputs x_{t-1}, x_t, and x_{t+1}, respectively. At step t, the operation that occurs in the hidden layer can be expressed as

$$h^{(t)} = f\left(h^{(t-1)}, x^{(t)}; W\right)$$

The expansion, or unfolding, thus done over time leads to a network that now looks similar to a feedforward network with multiple hidden layers, each representing the operations conducted at one time step. This expansion depends on sequence length and how many such combinations, given by batch size.

You can think of unfolded structure as a feedforward network. During the training phase, our aim is to learn the weights by computing the loss based on the prediction with respect to the actual output label and accordingly compute the gradients for backpropagation, which will adjust the weights to produce a better prediction in the next iteration.

If we unroll all input time steps, we extend the network to a structure with inputs at each time step leading to the output via hidden layers with common weights. Each time step in this technique can be seen as an extension of a hidden layer, with the previous time step's internal state serving as an input. Each time step has one input time step, one copy of the network, and one output. Errors are then calculated and accumulated for each time step. The network is rolled back up, and the weights are updated. This is called backpropagation through time (BPTT).

We'll observe that due to the chain rule, we will often see similar multiplication operations on very small values happening over and over again, which may cause the result to approach 0 or infinity. We will discuss this issue and a common potential solution later.

Types of RNN

Sequence problems are not the same. However, the concepts behind RNN provide us ample ideas to create a recurrent neural network for solving various problems. Depending on the situation, RNNs can take different forms, which determine the type of RNN to be used for a particular problem. Figure 15-4 shows a simplified structure of different types of RNNs discussed in the following.

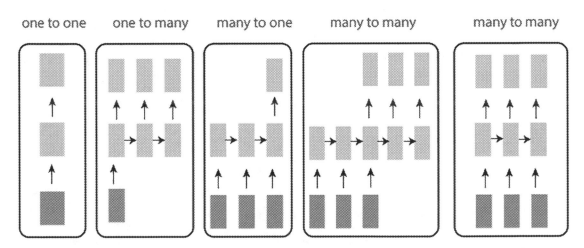

Figure 15-4. *Types of recurrent neural networks*

One to One

One to one is the form that doesn't require recurrent units, and each input is used by a hidden layer, followed by the output layer. These are the plain neural networks that deal with a fixed size of the input to the fixed size of the output, where they are independent of previous information/output.

One to Many

These recurrent neural networks take a fixed size of input and produce a sequence of data as output. These are used in solving problems like image captioning and music generation. The input is a static size, for example, an image, and the output is a sequence of words.

Many to One

These recurrent neural networks take a sequence of information as input and produce output of a fixed size. Thus, input should be either a sequence of words or an audio signal, which may yield a fixed size output like sentiment polarity (positive or negative label).

Many to Many

Many-to-many RNNs are of two types. The first type is the RNNs in which the input size is equal to the output size – thus in a manner, one output being produced for each item of input sequence. One popular example is sequence labelling, or entity recognition. For each word in a sentence, we generate labels like a Person Name, Location, Organization, or any Other word. The other type of RNNs may produce output sequence of a different size compared to the input sequence. These are highly helpful in machine translation, where each word in one language may not be directly mapped to a word in the second language, but how the words appear in a sequence produces a meaning in the target language.

Thus, based on the configuration, recurrent neural networks allow processing inputs of arbitrary length and generate the outputs accordingly. However, a major drawback they face is that at any time step, the network leverages the current input and the past outputs but can't make any changes based on a future input. However, for most applications, it is enough to build a model using RNN to capture the historic information. We'll revisit such scenarios in a future section.

RNN in Python

PyTorch provides an RNN implementation that requires the specifications of your recurrent layer and can be added as part of a larger network. It applies the weights and bias to the input data and combines that with the weighted hidden statuses and finally applies a nonlinear function (ReLU or tanH) to that. This provides the hidden state for the next input.

Let's begin with a simple example that takes a sequence of numbers. Let's begin with importing the required libraries. RNN is defined under `torch.nn`.

```python
import torch
import torch.nn as nn
import torch.optim as optim
import numpy as np
from torch.utils.data import Dataset, DataLoader
```

Let's create a simple sequence as a tensor.

```python
data = torch.Tensor([1, 2, 3, 4, 5, 6, 7, 8, 9, 10, 11, 12, 13, 14, 15, 16,
17, 18, 19, 20])
print("Data: ", data.shape, "\n\n", data)
```

```
Out:
Data:  torch.Size([20])
tensor([ 1.,  2.,  3.,  4.,  5.,  6.,  7.,  8.,  9., 10., 11., 12., 13.,
14., 15., 16., 17., 18., 19., 20.])
```

To define a simple RNN, we need to define sequence length, batch size, and input size. In this example, we will predict the next two numbers based on five contiguous numbers of the sequence.

We have simple data with only one column, and the problem we're trying to solve says that we have to observe a sequence of length of five inputs to predict the next two outputs.

```python
INPUT_SIZE = 1
SEQ_LENGTH = 5
HIDDEN_SIZE = 2
NUM_LAYERS = 1
BATCH_SIZE = 4
```

In most cases, one or more layers of RNN will be part of a larger network that ends with an output layer. Based on the hyperparameters that specify how the layer in the network should look like, we can define the RNN layer using

```
rnn = nn.RNN(input_size=INPUT_SIZE, hidden_size=HIDDEN_SIZE, num_layers =
1, batch_first=True)
```

We can use the data as input and obtain the output and the hidden state using the following. The size of the input depends on batch size, sequence length, and input size – the parameters we defined previously. nn.RNN() can also accept pre-set values for hidden layer, which default to all zeros if not specified.

```
inputs = data.view(BATCH_SIZE, SEQ_LENGTH, INPUT_SIZE)
out, h_n = rnn(inputs)
```

The output we thus receive is the tensor containing output of the RNN from all time steps from the last RNN layer. It's size is (sequence length, batch, num_directions * hidden size), where num_directions is 2 for bidirectional RNNs, otherwise 1.

Long Short-Term Memory

As we saw in the previous sections, as the network grows larger, the RNNs are not able to successfully propagate the gradients that tend to become zero (or in some cases, infinite), thus leading to vanishing gradient problem. This problem is evident in problems where we receive a long sequence as input, for example, long sentences from a product review or a blog post.

Consider the following example:

> *I didn't go to Amsterdam this year but I was there during the Summer last year and it was quite crowded everywhere in the city.*

Not all the words are relevant to interpret the meaning of the sentence. But when we look at the whole sentence, it is important to know that the word "city" refers to the same entity that is referred to by the word Amsterdam. Long dependencies, as the ones shown in Figure 15-5, should be preserved.

In RNN, the input at the current time step and the hidden state computed in the previous time step are combined and passed through tanh activation. The output thus created also acts as the hidden state for the next time step. Long short-term memory, or

LSTM, is an architecture that partially solves the vanishing gradient problem by allowing the gradients to flow unchanged. LSTM cells are constructed specifically to remember the parts of the sequence that matter, thus keeping only the relevant information to make prediction.

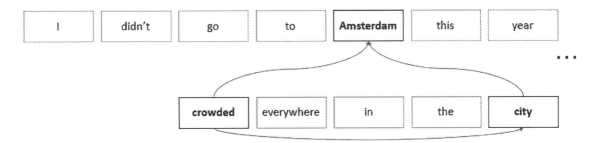

Figure 15-5. *A long sentence may connect long-term dependencies*

LSTMs are designed to hold a cell state that can be referred to as a memory. Due to a cell holding a memory, information from earlier time steps can be transferred to the later time steps, thus preserving long-term dependencies in sequences. The memory block is controlled by gates – which are simple operations that decide which information should be kept or forgotten during the training. This is implemented by applying sigmoid activation at certain computations within the cell that outputs zero for any negative values provided to it, thus leading to the concept of forgetting the information.

LSTM Cell

The structure of an LSTM Cell is shown in Figure 15-6. The first section of LSTM cell that receives the input is called a forget gate. **Forget gate** decides the information that should be kept or forgotten (or discarded). This gate combines the hidden state from the previous time step and the current input and passes them through the sigmoid function. Regardless of the forget operation, the input combined with the previous hidden state is passed through sigmoid function and a tanh function. The output of tanh function acts as the candidate, which may be propagated further, and the output of the sigmoid function acts as an evaluation function that decides the importance of the values. The tanh output is then multiplied with the sigmoid output. This section of the LSTM cell is called the **input gate**.

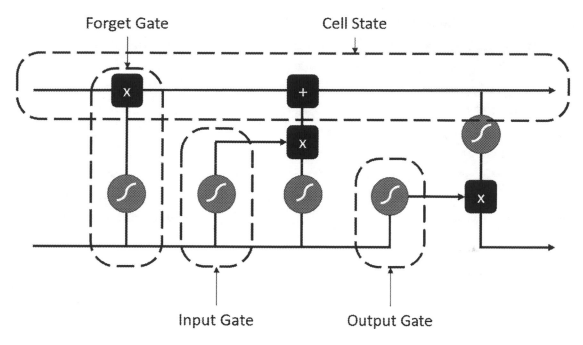

Figure 15-6. *Structure of a recurrent neural network cell*

The output of forget gate is a vector, which is multiplied with the previous cell state, thus making the values that should be forgotten zero. We add this to the vector computed by combining the value obtained at the input gate, thus providing a new cell state.

The output of the LSTM cell becomes the hidden state, which will be later combined with the input in the next step – meanwhile, the cell state gets propagated without further transformations to the next steps. The **output gate** combines the candidate (previous hidden state and current input, followed by a sigmoid activation) and the cell state that was just computed, on which a tanh activation is first applied by performing a multiplication that forms the next hidden state.

Time Series Prediction

In this example, we will use one of the data sources available publicly that contain information about statistics related to the COVID-19 outbreak. The COVID-19 pandemic is a global pandemic (currently ongoing, at the time of writing this book) that is caused by severe acute respiratory syndrome coronavirus 2. First identified in December 2019 in Wuhan (China), it quickly spread around the world causing millions of deaths and hospitalizations.

The COVID-19 outbreak also gave rise to the use of analytics in public healthcare, and a prominent use case that became widespread was predicting future growth in coronavirus cases so that the healthcare system can be planned accordingly. In this exercise, we will use a database containing COVID statistics in terms of cases and death in each country. The details are specified per date – thus leading to the possibility of modelling and learning sequential changes.

The data used for this example was obtained from the website of the **European Centre for Disease Prevention and Control**,[1] although you can use relatively newer data from any reliable sources. You may download the CSV file on their webpage or use wget command (on Linux) in the terminal or Jupyter Notebook. In Jupyter Notebook, you will need to begin the command with ! sign.

```
!wget https://opendata.ecdc.europa.eu/covid19/casedistribution/csv/data.csv
```

If wget is not installed on your system, you will receive an error message saying

```
'wget' is not recognized as an internal or external command,
operable program or batch file.
```

In that case, you may either install wget or download the CSV file from the website and move it to a location that is easily accessible from Jupyter Notebook. Once you have the data, import the requirements:

```
import numpy as np
import matplotlib.pyplot as plt
import pandas as pd
import torch
import torch.nn as nn
from torch.autograd import Variable
from sklearn.preprocessing import MinMaxScaler
```

You can load the data as Pandas dataframe using read_csv():

```
covid_data = pd.read_csv("data.csv")
covid_data
```

---

[1] www.ecdc.europa.eu/en/publications-data/
download-todays-data-geographic-distribution-covid-19-cases-worldwide

The dataset contains 61900 rows and 12 columns. The dataset contains 61900 rows and 12 columns. A screenshot of the dataframe is shown in Figure 15-7. Please note this might change if the dataset at this location is updated in the future.

	dateRep	day	month	year	cases	deaths	countriesAndTerritories	geoId	countryterritoryCode	popData2019	continentExp	Cumulativ
0	14/12/2020	14	12	2020	746	6	Afghanistan	AF	AFG	38041757.0	Asia	
1	13/12/2020	13	12	2020	298	9	Afghanistan	AF	AFG	38041757.0	Asia	
2	12/12/2020	12	12	2020	113	11	Afghanistan	AF	AFG	38041757.0	Asia	
3	11/12/2020	11	12	2020	63	10	Afghanistan	AF	AFG	38041757.0	Asia	
4	10/12/2020	10	12	2020	202	16	Afghanistan	AF	AFG	38041757.0	Asia	
...	...	...	...	...	...	...		...	...	...	...	
61895	25/03/2020	25	3	2020	0	0	Zimbabwe	ZW	ZWE	14645473.0	Africa	
61896	24/03/2020	24	3	2020	0	1	Zimbabwe	ZW	ZWE	14645473.0	Africa	
61897	23/03/2020	23	3	2020	0	0	Zimbabwe	ZW	ZWE	14645473.0	Africa	

Figure 15-7. *Dataset verified after loading in a Jupyter Notebook*

To limit this use case, we will try to model data about cases in one country. To filter the data, we will select only the rows containing countryterritoryCode as USA.

```
covid_data[covid_data['countryterritoryCode']=='USA']
```

After this, we need to index only the date column, dateRep, and cases through the following line:

```
data = covid_data[covid_data['countryterritoryCode']=='U
SA']      [['dateRep', 'cases']]
```

The data now contains only two columns from the country that was filtered. As date formats are different around the world, and the fact that the date has been loaded as object similar to other fields rather than a date/time field, we need to process it further.

```
data['dateRep'] = pd.to_datetime(data['dateRep'], format="%d/%m/%Y")
```

The data object can be sorted by date in ascending order so that we have starting dates at the top where cases were mostly 0. We will also convert date column to index.

```
data = data.sort_values(by="dateRep", key=pd.to_datetime)
data = data.set_index('dateRep')
```

The data object should now look similar to the screenshot shown in Figure 15-8.

dateRep	cases
2019-12-31	0
2020-01-01	0
2020-01-02	0
2020-01-03	0
2020-01-04	0
...	...
2020-12-10	220025
2020-12-11	224680
2020-12-12	234633
2020-12-13	216017
2020-12-14	189723

Figure 15-8. *Part of dataframe showing the growth of COVID-19 cases in the United States*

Let's quickly visualize how the cases grew in the United States in 2020.

```
fig = plt.gcf().set_size_inches(12,8)
plt.plot(data, label = 'Covid-19 Cases in the US')
plt.show()
```

Figure 15-9 shows that initially the numbers were close to zero but by April, the numbers started increasing and rose beyond 200,000 by the end of the year. In sequence modelling, we assume that the number on a particular day can be predicted based on how the numbers of the previous few days changed.

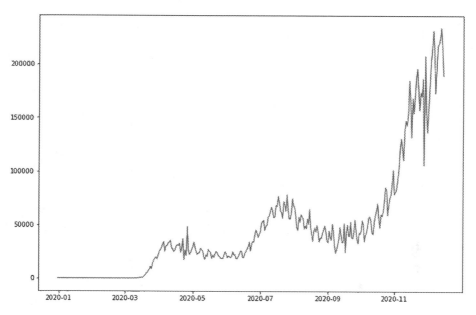

Figure 15-9. *Plot of COVID-19 cases in the United States*

As we have seen before, we need to find chunks of sequences. Say, the data is from a certain week starting from Sunday. In textual data, n-grams, most commonly bigrams and trigrams, are highly popular, though recently character chunks and character n-grams have also been suitable to problems that can be solved using RNN/LSTMs.

We will create a function that accepts a NumPy array and an integer defining the sequence length and return the chunks of mentioned sequence length. For example, [1,2,3,4,5,6] of sequence length 3 should create chunks as [1,2,3], [2,3,4], [3,4,5], and [4,5,6]. However, in this problem, we also want to predict the next element; that is, 1,2,3 should be followed by 4. The chunks will look slightly different as shown in Figure 15-10.

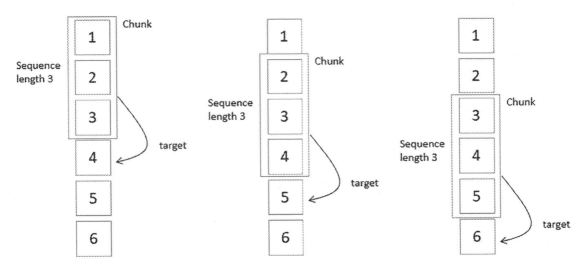

Figure 15-10. *Window or chunks of a long sequence to predict the next item*

Here's one of the possible implementations.

```
def chunkify(data, seq_length):
    chunks = []
    targets = []

    for i in range(len(data)-seq_length-1):
        chunks.append(data[i:(i+seq_length)])
        targets.append(data[i+seq_length])

    return np.array(chunks),np.array(targets)
```

Each number in the sequence is a data element; thus, you can reshape the array as nx1 shape. Thus, we can proceed to call

```
chunkify([[1],[2],[3],[4],[5],[6]],3)
Out:
(array([[[1],
         [2],
         [3]],

        [[2],
         [3],
```

```
        [4]]]),
 array([[4],
        [5]]))
```

Preprocessing of the data would also require to scale the data to the right values. We can use MinMaxScaler. We will use 80% of the data as training set and keep a sequence length of 5.

```
sc = MinMaxScaler()
training_data = sc.fit_transform(data.values.copy())

seq_length = 5
x, y = chunkify(training_data, seq_length)

train_size = int(len(y) * 0.8)
test_size = len(y) - train_size
```

Before going further, we need to convert the datasets into tensors. We can convert the chunks and targets up to index train_size to use them with PyTorch network.

```
dataX = Variable(torch.Tensor(np.array(x)))
dataY = Variable(torch.Tensor(np.array(y)))

trainX = Variable(torch.Tensor(np.array(x[0:train_size])))
trainY = Variable(torch.Tensor(np.array(y[0:train_size])))

testX = Variable(torch.Tensor(np.array(x[train_size:len(x)])))
testY = Variable(torch.Tensor(np.array(y[train_size:len(y)])))
```

In the network, we will define an LSTM layer followed by a fully connected layer. In PyTorch, LSTM is defined in torch.nn. It accepts the parameters like input size, hidden state feature size, and number of recurrent layers. If num_layers is specified, the model will stack that number of LSTMs together so that the output of the first LSTM is provided to the second LSTM, and so on. As this is a regression problem, criteria can be mean squared error, which is used widely in regression problems.

Let's create a model class. We will need to define the individual layers. We will keep the structure simple and have one LSTM layer, followed by fully connected output. In forward() method, we will start by creating a hidden state and cell memory initialized with zeros. The output of LSTM will be applied to the fully connected layer and returned as output.

```python
class CovidPrediction(nn.Module):

    def __init__(self, num_classes, input_size, hidden_size, num_layers):
        super(CovidPrediction, self).__init__()

        self.num_classes = num_classes
        self.num_layers = num_layers
        self.input_size = input_size
        self.hidden_size = hidden_size
        self.seq_length = seq_length

        self.lstm = nn.LSTM(input_size=input_size, hidden_size=hidden_size,
        num_layers=num_layers, batch_first=True)

        self.fully_connected = nn.Linear(hidden_size, num_classes)

    def forward(self, x):
        h0 = Variable(torch.zeros(self.num_layers, x.size(0),
        self.hidden_size))
        c0 = Variable(torch.zeros(self.num_layers, x.size(0), self.
        hidden_size))
        ula, (h_out, _) = self.lstm(x, (h0, c0))
        h_out = h_out.view(-1, self.hidden_size)
        out = self.fully_connected(h_out)
        return out
```

We can see the structure of the model here:

```python
model = CovidPrediction(1, 1, 4, 1)
print(model)
Out:
CovidPrediction(
  (lstm): LSTM(1, 4, batch_first=True)
  (fully_connected): Linear(in_features=4, out_features=1, bias=True)
)
```

The model is fairly simple. Let's defined the criterion and optimizer and run the training loop.

```
num_epochs = 1000
learning_rate = 0.01

model = CovidPrediction(1, 1, 4, 1)


criterion = torch.nn.MSELoss()
optimizer = torch.optim.Adam(model.parameters(), lr=learning_rate)

for epoch in range(num_epochs):
    outputs = model(trainX)
    loss = criterion(outputs, trainY)
    loss.backward()

    optimizer.step()
    if epoch % 100 == 0:
      print("Iteration: %d, loss:%f" % (epoch, loss.item()))
```

The training loop will show the change in losses over time.

```
Iteration: 0, loss:0.162355
Iteration: 1, loss:0.162353
Iteration: 2, loss:0.162353
Iteration: 3, loss:0.162353
Iteration: 4, loss:0.162353
```

Once the training is complete, you can predict a target that should follow each chunk of five values. In other words, by looking at the number of COVID-19 infections of five consecutive days, we're going to predict the number of infections expected on the sixth day.

To visualize how closely the model has predicted the future cases, let's predict for all the possible chunks in the training dataset and compare them with the actual values in the training dataset.

```
model.eval()
train_predict = model(dataX)
```

We will need to convert the values into NumPy arrays.

```
data_predict = train_predict.data.numpy()
data_actual = dataY.data.numpy()
```

As the data was earlier scaled using MinMax transformation, we need to apply inverse transformation to get the actual values.

```
data_predict = sc.inverse_transform(data_predict)
data_actual = sc.inverse_transform(data_actual)
```

Let's visualize and compare. Figure 15-11 shows the output of the following code. You can see that in the later part, model results deviated from the actual because of limits of values in the training data.

```
fig = plt.gcf().set_size_inches(12,8)
plt.plot(data_actual)
plt.plot(data_predict)
plt.suptitle('')
plt.legend(['Actual cases in 2020', 'Predicted cases (latter 20% not in
training) '], loc='upper left')
```

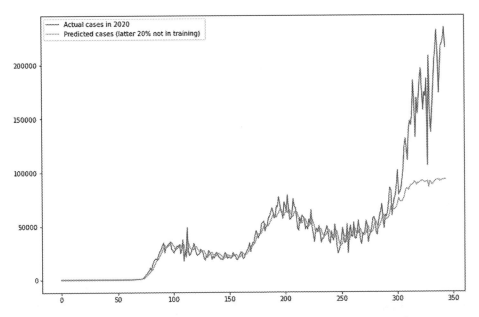

Figure 15-11. *Actual and predicted COVID-19 cases in the United States*

Though the model looks reliable for most of the data, due to the nature of the data (COVID cases had still been increasing), the model is not able to capture these spikes because (1) scaling would limit the model's understanding of very high values and (2) there were not enough spike patterns present in the training data.

In a subsequent run, we trained for 5000 epochs on 100% of the US training data and applied the model on COVID-19 statistics from India, and this led to a close prediction.

```
data = covid_data[covid_data['countryterritoryCode']=='IND'] [['dateRep',
'cases']]
data['dateRep'] = pd.to_datetime(data['dateRep'], format="%d/%m/%Y")
data = data.sort_values(by="dateRep", key=pd.to_datetime)
data = data.set_index('dateRep')
```

For prediction, we will modify the relevant sections in the prediction block.

```
sc = MinMaxScaler()
seq_length = 5
training_data = sc.fit_transform(data.values.copy())
x, y = chunkify(training_data, seq_length)
dataX = Variable(torch.Tensor(np.array(x)))
dataY = Variable(torch.Tensor(np.array(y)))
```

We will use the same block for evaluation.

```
model.eval()
train_predict = model(dataX)

data_predict = train_predict.data.numpy()
data_actual = dataY.data.numpy()

data_predict = sc.inverse_transform(data_predict)
data_actual = sc.inverse_transform(data_actual)

# plt.axvline(x=train_size, c='r', linestyle='--')
fig = plt.gcf().set_size_inches(12,8)
plt.plot(data_actual)
plt.plot(data_predict)
plt.suptitle('')
plt.legend(['Actual Covid-19 cases in India in 2020', 'Predicted cases of
Covid-19'], loc='upper left')
```

The plot comparing the actual and predicted number of cases in India is shown in Figure 15-12. You can see that the model is able to closely predict COVID-19 cases. We'd recommend you to try building a similar model for other sequential quantities as well as text at character n-gram levels.

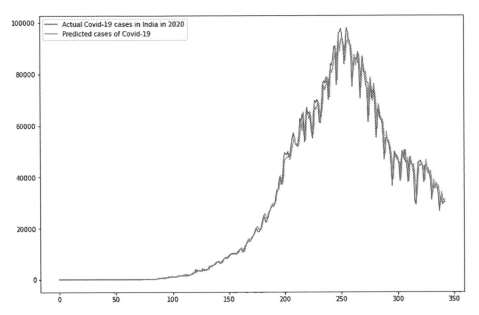

Figure 15-12. *Actual and predicted COVID-19 cases in India*

Gated Recurrent Unit

RNNs and LSTMs have become highly popular in the previous decade, which have given rise to the popularity of more sequence modelling architectures. One such architecture is a gated recurrent unit (GRU) introduced by Cho et al. in 2014. The structure of a GRU is shown in Figure 15-13.

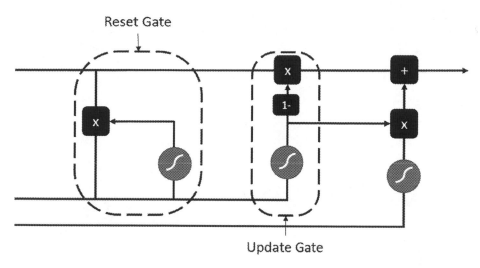

Figure 15-13. *Structure of a gated recurrent unit cell*

GRU instead introduced an Update Gate that decides which new information should be added or discarded and a Reset Gate that discards part of the past information. These are similar to LSTM but lack an output gate and use lesser parameters, thus relatively faster to train. These are highly popular in music modelling, speech modelling, and NLP. PyTorch provides implementation of GRU under nn.GRU().

Summary

In this chapter, we learned about recurrent neural networks and LSTMs and saw how they help in capturing the sequential nature of the data and, with some modifications, preserve the relationships through a sequence. The past few chapters have provided us few more sophisticated tools that have been used to build the state of the art in the field of machine learning and artificial intelligence.

In the next chapter, we will wrap it all together and discuss how data science and artificial intelligence projects are structured, planned, implemented, and deployed.

CHAPTER 16

Bringing It All Together

The past chapters in this book have introduced data analysis methods, feature extraction techniques, and traditional machine learning and deep learning techniques. We have conducted multiple experiments on numeric, textual, and visual data and found how to analyze and tweak the performance.

When you are working as a part of a large team trying to solve a business or a research problem, or building a complex AI-powered software that will be used by millions of users, you have to plan the project with the end goal in mind (Figure 16-1). This brings you to consider the management and engineering side of data science.

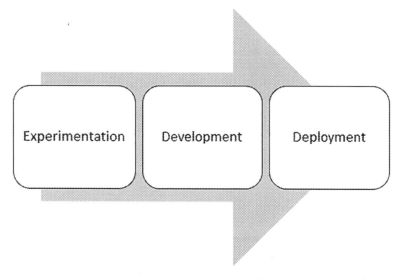

Figure 16-1. *Practices for successful deployment of machine learning projects*

In this chapter, we're going to discuss strategies for planning data science and artificial intelligence projects, tools for persisting the models, and hosting the models as a microservice that can be used in the evolving applications.

© Ashwin Pajankar and Aditya Joshi 2022
A. Pajankar and A. Joshi, *Hands-on Machine Learning with Python*, https://doi.org/10.1007/978-1-4842-7921-2_16

Data Science Life Cycle

Data science and artificial intelligence projects are complex, and it is very easy to get caught up in smaller details or focus too much on creating models and hosting them while losing the sight of long-term vision. Every data science project is different and might be managed using different frameworks and processes – however, all the projects have similar steps (Figure 16-2).

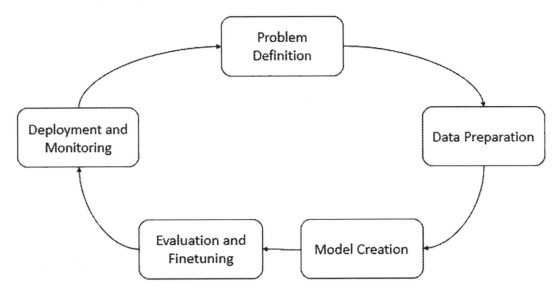

Figure 16-2. *Iterative data science life cyle process*

The process usually begins with focus on defining the business or research objectives and coming up with the artifacts that properly define the problem we are trying to solve. This leads to a clear understanding about the data that will be required, which then expands to analysis of data sources, technical expertise and cost required to obtain the data, and evaluation of data in terms of how nicely will it support in reaching the business objective. Once the data has been obtained, we might need to clean, preprocess, and, in some cases, combine multiple data sources to enrich the quality of data.

Next step in the process is model creation. Based on the business objectives and technological constraints, we decide what kind of solutions might be applicable to this problem. We often begin with simple experiments with basic feature engineering and out-of-the-box solutions and then proceed to more thorough model developments. Based on the type of data, chosen solution, and availability of computation power, this can take hours to days to development as well as training. This is closely tied with thorough evaluation and tuning.

This life cycle is not a rigid structure but shows the process at a top level. The aim of such processes is to provide a standard set of steps involved, along with details about information required for each such step, and the deliverables and documentations that are produced. One such highly popular framework is CRISP-DM.

CRISP-DM Process

CRoss **I**ndustry **S**tructured **P**rocess for **D**ata **M**ining is a process framework that defines the common tasks in data-intensive projects that are done in a series of phases with an aim to create repeatable processes for data mining applications. It is an open standard that is developed and followed by hundreds of large enterprises around the world. It was originally devised in 1996, which led to the creation of CRISP-DM Special Interest group that obtained funding from the European Commission and led to a series of workshops that over the past decades have been defining and refining the process and artifacts involved.

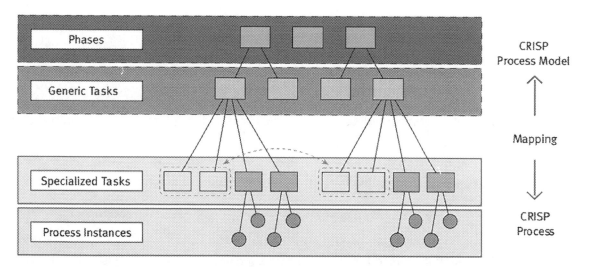

Figure 16-3. *CRISP-DM methodology*

Figure 16-3 shows how the process model is designed at four levels of abstraction. At the top level, the phases define several generic tasks that are meant to be well-defined, complete, and stable tasks, which are carried out as special tasks. There may be a generic task called **Collect User Data**, which may require specialized tasks like (1) export users table from the database, (2) find user location using external service, and (3) download data from user's LinkedIn profile using the API. The fourth level covers the actual implementation of the specialized tasks – also covering a record of actions, decision, and results of the task that is performed.

There are six phases of the CRISP-DM model. The following sections describe each one.

Phase 1: Business Understanding

Before diving deeper into the project, the first step is to understand the end goal of the project from the stakeholder's point of view. There might be conflicting objectives, which, if not analyzed at this level, may lead to unnecessary repetition costs. By the end of this phase, we will have a clear set of business objectives and business success criteria. We also conduct analysis of resources availability and risk during assess situation. After this, we then define the goals of the project from a technical data mining perspective and produce a project plan.

Phase 2: Data Understanding

This phase involves tasks for collecting initial data. Most projects require data from multiple sources which need to be integrated – that can be covered either in this phase of the next. However, the important part here is to create an initial data collection report that explains how the data was acquired and what problems were encountered. This phase also covers data exploration and describing the data along with verifying data quality. Any potential data quality issues must be addressed.

Phase 3: Data Preparation

The data preparation phase assumes that initial data has been obtained and studied and potential risks have been planned. The end goal of this goal is to produce the ready-to-use datasets that will be used for modelling or analysis. An additional artifact will describe the dataset.

As a part of this phase, select the datasets – and for each dataset, document the reasons for inclusion and exclusion. This is followed by data cleaning, in which the data quality is improved. This may involve transformation, deriving more attributes or enriching the datasets. After cleaning, transformation, and integration, the data is formatted to make it simpler to load the data in the future stages.

Phase 4: Modelling

Modelling is the phase in which you build and assess various models based on the different modelling and machine learning techniques we have studied so far. At the first step, the modelling technique to be used is selected. There will be different instances of this task based on different modelling methods or algorithms that you wish to explore and evaluate. You will generate a test design, build the model, assess it thoroughly, and evaluate how closely the model fits the technical needs of the system.

Phase 5: Evaluation

Evaluation phase looks broadly at which model meets the business needs. The tasks involved in this phase test the models in real application and assess the results generated. After this, there are tasks on Review process, in which we do a thorough review of the data mining engagement in order to determine if there is any important factor or task that should have been covered. Finally, we determine the next steps to decide whether the models require further tuning or move to the deployment of the model. At the end of this phase, we have documented the quality of the models and a list of possible actions that should be taken next.

Phase 6: Deployment

The final phase, deployment, is the phase that brings the work done so far to the actual use. This phase varies widely based on the business needs, organization policies, and engineering needs. This begins with planning deployment, involves developing a deployment plan containing the strategy for deployment. We also need to plan a thorough monitoring and maintenance plan to avoid issues after the end-to-end project has been launched. Finally, the project team documents a summary of the project and conducts a project review to discuss and document what went well, what could have been better, and how to improve in the future.

In practice, most organizations use these phases as guidelines and create their own processes based on their budgets, governance requirements, and needs. Many small-scale teams might not follow these steps and get captured in a long loop of iterations and iterations of development and improvements, not being able to avoid the pitfalls that otherwise could have been well planned and handled if these were studied.

In the next part of this chapter, we will study the technical aspects of development and deployment of data science and AI projects.

How ML Applications Are Served

Once a model has been created, it has to be integrated with larger enterprise application. A most common form of serving the models is as a service or a microservice. The aim of this kind of architecture (Figure 16-4) is to encapsulate the whole workflow of prediction/inference process including data preparation, feature extraction, loading a previously created model, predicting the output values, and, often, logging together through an easy-to-use interface. These interfaces are most commonly served as an endpoint in a web server.

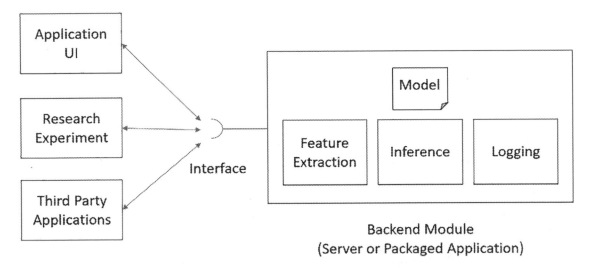

Figure 16-4. *Serving ML models as a microservice*

In larger applications, these servers are hosted on cloud often through Docker for easy deployment. The concept of deploying, monitoring, and maintaining machine learning models for AI applications is being expanded into well-structured concepts in the form of MLOps.

In the next few pages, we will take a small project that will be eventually hosted as an ML application.

Learning with an Example

In this mini-project, we will build a sentiment analysis tool using PyTorch with an aim to experiment with model architecture to achieve relatively good performance, save the parameters, and host it using flask.

The first attempt toward sentiment analysis was General Inquirer system, published in 1961. The typical task in sentiment analysis is text polarity classification, where the classes of interest are positive and negative, sometimes with a neutral class. With advancement in computational capabilities, machine learning algorithms, and later deep learning, sentiment analysis is much more accurate and prevalent in a lot of situations.

Defining the Problem

Sentiment analysis is a vast field that covers the problem of identifying emotions, opinions, moods, and attitudes. There are also many names and slightly different tasks, for example, sentiment analysis, opinion mining, opinion extraction, sentiment mining, subjectivity analysis, affect analysis, emotion analysis, review mining, etc.

In this problem, we will build a model for classifying whether a movie review sentence is positive, negative, or neutral. In traditional machine learning approaches, feature engineering would be the primary task. A feature vector is a representation of actual content (document, tweet, etc.) that the classification algorithm takes as an input. The purpose of a feature, other than being an attribute, would be much easier to understand in the context of a problem. A feature is a characteristic that might help when solving the problem.

In a deep learning solution, we can either use embeddings or sequence of characters. But first, we have to obtain the data.

In some cases, you will collect the data through your database logs, or hire a data gathering team, or like in our case, get lucky and stumble over a freely available dataset. A 50,000-item movie review dataset[1] has been gathered and prepared by Stanford, published in 2011.

Data

You can download the data from their webpage though the solutions we're going to explain here will work equally fine with other datasets including the ones from product reviews or social media text. The dataset downloaded from the website contains a compressed tar file, which after decompression expands into two folders, namely, test and train, and some additional files containing information about the dataset. An

---

[1] http://ai.stanford.edu/~amaas/data/sentiment/

alternate copy of the dataset that has been preprocessed is available on Kaggle,[2] which is shared by Lakshmipathi N.

The dataset contains 50,000 reviews, each of which is marked as positive or negative. This gives an indication about the last layer of the neural network structure – all we need is a single node with sigmoid activation function. If there were more than two classes, say, positive, negative, or neutral, we would create three nodes, each representing a sentiment class label. The node with the highest value would indicate the predicted result.

Assuming that you have downloaded the dataset containing one review and sentiment per row in a CSV, we can start exploring it.

```
import pandas as pd
dataset = "data/IMDB Dataset.csv"
df = pd.read_csv(dataset, sep=",")
sample_row = df.sample()
sample_row['review'].values
```

Out:
```
array(["Etienne Girardot is just a character actor--the sort of person
people almost never would know by name. However, he once again plays the
coroner--one of the only actors in the Philo Vance films that played his
role more than once.
```

The output has been truncated here for brevity. You can see the labelled sentiment for this review using `sample_row['sentiment']`, which is `positive` for this sample.

To prepare for a machine learning experiment, we'll divide the data into training and testing datasets.

```
from sklearn.model_selection import train_test_split
X,y = df['review'].values,df['sentiment'].values
X_traindata,X_testdata,y_traindata,y_testdata = train_test_
split(X,y,stratify=y)

print(f'Training Data shape : {X_traindata.shape}')
print(f'Test Data shape     : {X_testdata.shape}')
```

---

[2] www.kaggle.com/lakshmi25npathi/imdb-dataset-of-50k-movie-reviews

```
print(f'Training Target shape : {y_traindata.shape}')
print(f'Test Target shape     : {y_testdata.shape}')
Out:
Training Data shape : (37500,)x
Test Data shape     : (12500,)
Training Target shape : (37500,)
Test Target shape     : (12500,)
```

We know that most models require the data to be converted to a particular format. In our RNN-based model, we will need to convert the data into a sequence of numbers – where each number represents a word in the vocabulary.

In preprocessing stage, we will need to (1) convert all the words to lowercase, (2) tokenize and clean the string, (3) remove stop words, and (4) based on our knowledge of words in the training corpus, prepare a dictionary of words and convert all the words to numbers based on the dictionary.

We will also convert the sentiment labels so that negative is represented by a 0 and positive by a 1. In this implementation, we are limiting the vocabulary size to 2000 – thus, the most frequent 2000 words will be considered while creating the sequences and others will be ignored. You can experiment with changing this number based on computational capacity and the target quality of results. The implementation is shown here:

```
import re
import numPy as np
from collections import Counter

def preprocess_string(s):
    s = re.sub(r"[^\w\s]", '', s)
    s = re.sub(r"\s+", '', s)
    s = re.sub(r"\d", '', s)
    return s

def mytokenizer(x_train,y_train,x_val,y_val):
    word_list = []

    stop_words = set(stopwords.words('english'))
    for sent in x_train:
        for word in sent.lower().split():
```

```
        word = preprocess_string(word)
        if word not in stop_words and word != '':
            word_list.append(word)

    corpus = Counter(word_list)
    corpus_ = sorted(corpus,key=corpus.get,reverse=True)[:2000]
    onehot_dict = {w:i+1 for i,w in enumerate(corpus_)}

    final_list_train,final_list_test = [],[]
    for sent in x_train:
            final_list_train.append([onehot_dict[preprocess_string(word)]
            for word in sent.lower().split() if preprocess_string(word) in
            onehot_dict.keys()])
    for sent in x_val:
            final_list_test.append([onehot_dict[preprocess_string(word)]
            for word in sent.lower().split() if preprocess_string(word) in
            onehot_dict.keys()])

    encoded_train = [1 if label =='positive' else 0 for label in y_train]
    encoded_test = [1 if label =='positive' else 0 for label in y_val]
    return np.array(final_list_train), np.array(encoded_train),np.
    array(final_list_test), np.array(encoded_test),onehot_dict
```

We can now prepare the training and test arrays using

```
X_train,y_train,X_test,y_test,vocab = mytokenizer(X_traindata,y_
traindata,X_testdata,y_testdata)
```

There's a possibility that you might get an error message that looks like this:

```
LookupError:
*******************************************************
  Resource stopwords not found.
  Please use the NLTK Downloader to obtain the resource:

  >>> import nltk
  >>> nltk.download('stopwords')

  For more information see: https://www.nltk.org/data.html
  Attempted to load corpora/stopwords
```

```
Searched in:
  - 'C:\\Users\\JohnDoe/nltk_data'
  - 'C:\\Users\\ JohnDoe \\Anaconda3\\nltk_data'
  - 'C:\\Users\\ JohnDoe \\Anaconda3\\share\\nltk_data'
  - 'C:\\Users\\ JohnDoe \\Anaconda3\\lib\\nltk_data'
  - 'C:\\Users\\ JohnDoe \\AppData\\Roaming\\nltk_data'
  - 'C:\\nltk_data'
  - 'D:\\nltk_data'
  - 'E:\\nltk_data'
**********************************************************
```

This means we do not have a list of stopwords in nltk, which we can install using nltk. download(). This is required only once in your python environment. For more details, you can refer to the NLTK[3] documentation.

Alternatively, you can construct a list of stopwords and add the logic to remove the words present in the stopwords list.

Before proceeding further, we should verify if the objects are in the right shape and size.

vocab should be a dictionary of length 2000 (the number we limit the vocabulary size to). X_train, y_train, X_test, and y_test should be numpy.ndarray with the size same as the split of the original dataset.

As we are planning to use RNN for this task, RNNs use sequences to be of a certain length – thus, if a sentence is too short, we will have to pad the sequence, potentially with 0s. If a sentence is too long, we have to decide a maximum length and truncate – which might happen more often during the prediction phase when we encounter previously unseen data. To decide that length, let's explore the training dataset and see how long the reviews are.

```
review_length = [len(i) for i in X_train]
print ("Average Review Length : {} \nMaximum Review Length : {}
".format(pd.Series(review_length).mean(), pd.Series(review_length).max()))
Out:
Average Review Length : 81.74666666666667
Maximum Review Length : 662
```

---

[3] www.nltk.org/data.html

The review lengths look quite long by looking at these. However, in general, we'll have a lot of reviews that are short, and there will be very few that are very long. It will not be wrong to truncate review sequence length to 200 words, thus, of course, losing information in reviews that are more than 200 words long, but we assume they will be quite rare and should not have much impact on the performance of the model.

Each review, if long, should be limited to 200 words. What if it is shorter than 200 words? We will pad it with empty cells (or zeros).

```
def pad(sentences, seq_len):
    features = np.zeros((len(sentences), seq_len),dtype=int)
    for ii, review in enumerate(sentences):
        if len(review) != 0:
            features[ii, -len(review):] = np.array(review)[:seq_len]
    return features
```

We can test it before using with the training dataset. In the following lines, we'll pass a data row with ten elements, and the function will be called to pad it to make its length 20.

```
test = pad(np.array([list(range(10))]), 20)
test
Out: array([[0, 0, 0, 0, 0, 0, 0, 0, 0, 0, 0, 1, 2, 3, 4, 5, 6, 7, 8, 9]])
```

If we instead send numbers from 1 to 30, the function would truncate the ten trailing numbers.

```
test = pad(np.array([list(range(30))]), 20)
test
Out: array([[ 0,  1,  2,  3,  4,  5,  6,  7,  8,  9, 10, 11, 12, 13, 14,
15, 16, 17, 18, 19]])
```

Test and train dataset can be padded now.

```
X_train = pad(X_train,200)
X_test = pad(X_test,200)
```

The data is now ready. We can proceed to define the neural network. If we have a GPU available, we'll set our device to GPU. We'll use this variable later in our code.

```
import torch

is_cuda = torch.cuda.is_available()

if is_cuda:
    device = torch.device("cuda")
else:
    device = torch.device("cpu")
print (device)
Out:
    GPU
```

Preparing the Model

We will first need to convert the datasets into tensors. We can use `torch.from_numpy()` to create tensors, followed by `TensorDataset()` to pack the data values and labels together. We can then define `DataLoaders` to load the data for larger experiments.

```
import torch
import torch.nn as nn
import torch.nn.functional as F
from torch.utils.data import TensorDataset, DataLoader
# create Tensor datasets
train_data = TensorDataset(torch.from_numpy(X_train), torch.from_
numpy(y_train))
valid_data = TensorDataset(torch.from_numpy(X_test), torch.from_
numpy(y_test))

batch_size=50

train_loader = DataLoader(train_data, shuffle=True, batch_size=batch_size)
valid_loader = DataLoader(valid_data, shuffle=True, batch_size=batch_size)
```

The model is a simple model that will have an input layer, followed by the LSTM layer, followed by a one-unit output layer with sigmoid activation. We will use a dropout layer for basic regularization to avoid overfitting.

The input layer is an embedding layer with shape representing the batch size, layer size, and the sequence length. In the model class, the forward() method will implement the forward propagation computations. We will also implement init_hidden() method to initialize the hidden state of LSTM to zeros. The hidden state stores the internal state of the RNN from predictions made on previous tokens in the current sequence to maintain the concept of memory within the sequence. However, it is important that when we read the first token of the next review, the state should be reset, which will be updated and used by the rest of the tokens. The implementation is given here:

```
class SentimentAnalysisModel(nn.Module):
    def __init__(self, no_layers, vocab_size, hidden_dim, embedding_dim,
    drop_prob=0.5):
        super(SentimentAnalysisModel,self).__init__()

        self.output_dim = output_dim
        self.hidden_dim = hidden_dim

        self.no_layers = no_layers
        self.vocab_size = vocab_size

        self.embedding = nn.Embedding(vocab_size, embedding_dim)

        self.lstm = nn.LSTM(input_size=embedding_dim, hidden_size=self.
        hidden_dim, num_layers=no_layers, batch_first=True)
        self.dropout = nn.Dropout(0.3)

        self.fc = nn.Linear(self.hidden_dim, output_dim)
        self.sig = nn.Sigmoid()

    def forward(self,x,hidden):
        batch_size = x.size(0)
        embeds = self.embedding(x)
        lstm_out, hidden = self.lstm(embeds, hidden)
        lstm_out = lstm_out.contiguous().view(-1, self.hidden_dim)
        out = self.dropout(lstm_out)
        out = self.fc(out)
        sig_out = self.sig(out)
        sig_out = sig_out.view(batch_size, -1)
        sig_out = sig_out[:, -1] # get last batch of labels
```

```
        return sig_out, hidden

    def init_hidden(self, batch_size):
        ''' Initializes hidden state '''
        h0 = torch.zeros((self.no_layers, batch_size, self.hidden_dim)).
        to(device)
        c0 = torch.zeros((self.no_layers, batch_size, self.hidden_dim)).
        to(device)
        hidden = (h0,c0)
        return hidden
```

We can now initialize the model object. The hyperparameters can be defined separately for fine-tuning the model later. no_layers will be used to define stacking of RNNs. vocab_size is being incremented to adjust the shape of the embedding layer to accommodate 0s for padded reviews. output_dim is 1 to have only one node in the output layer that will contain a number that can be seen as probability of a review being positive. hidden_dim is used to specify the size of hidden state in the LSTM.

```
no_layers = 2
vocab_size = len(vocab) + 1
embedding_dim = 64
output_dim = 1
hidden_dim = 256

model = SentimentAnalysisModel (no_layers,  vocab_size, hidden_dim,
embedding_dim, drop_prob=0.5)
model.to(device)

print(model)
```

The model output should look like the following:

```
SentimentAnalysisModel(
  (embedding): Embedding(2001, 64)
  (lstm): LSTM(64, 256, num_layers=2, batch_first=True)
  (dropout): Dropout(p=0.3, inplace=False)
  (fc): Linear(in_features=256, out_features=1, bias=True)
  (sig): Sigmoid()
)
```

Let's define the training loop. We will use binary cross entropy loss function, which is a good choice for simple binary classification problems. We will keep learning rate as 0.01, and the optimizer is Adam optimization algorithm.

The training loop can be made to run for a large number of epochs. We will keep a track of accuracy and losses over each epoch to see how the performance improves over multiple iterations of training.

The accuracy will simply compare the output at the output layer and round it. Across all the examples, the accuracy will simply show the ratio of correctly labelled training data points.

```
def acc(pred,label):
    return torch.sum(torch.round(pred.squeeze()) == label.squeeze()).item()
```

The training loop is implemented as follows:

```
lr=0.001
criterion = nn.BCELoss()
optimizer = torch.optim.Adam(model.parameters(), lr=lr)

clip = 5
epochs = 10
valid_loss_min = np.Inf
# train for some number of epochs
epoch_tr_loss,epoch_vl_loss = [],[]
epoch_tr_acc,epoch_vl_acc = [],[]

for epoch in range(epochs):
    train_losses = []
    train_acc = 0.0
    model.train()
    # initialize hidden state
    h = model.init_hidden(batch_size)
    for inputs, labels in train_loader:

        inputs, labels = inputs.to(device), labels.to(device)
        # Creating new variables for the hidden state, otherwise
        # we'd backprop through the entire training history
        h = tuple([each.data for each in h])
```

```
        model.zero_grad()
        output,h = model(inputs,h)
        loss = criterion(output.squeeze(), labels.float())
        loss.backward()

        train_losses.append(loss.item())
        # calculating accuracy
        accuracy = acc(output,labels)
        train_acc += accuracy
        #`clip_grad_norm` helps prevent the exploding gradient problem in
        RNNs / LSTMs.
        nn.utils.clip_grad_norm_(model.parameters(), clip)
        optimizer.step()
val_h = model.init_hidden(batch_size)
val_losses = []
val_acc = 0.0
model.eval()
for inputs, labels in valid_loader:
        val_h = tuple([each.data for each in val_h])

        inputs, labels = inputs.to(device), labels.to(device)

        output, val_h = model(inputs, val_h)
        val_loss = criterion(output.squeeze(), labels.float())

        val_losses.append(val_loss.item())

        accuracy = acc(output,labels)
        val_acc += accuracy

epoch_train_loss = np.mean(train_losses)
epoch_val_loss = np.mean(val_losses)
epoch_train_acc = train_acc/len(train_loader.dataset)
epoch_val_acc = val_acc/len(valid_loader.dataset)
epoch_tr_loss.append(epoch_train_loss)
epoch_vl_loss.append(epoch_val_loss)
epoch_tr_acc.append(epoch_train_acc)
epoch_vl_acc.append(epoch_val_acc)
```

```
print(f'Epoch {epoch+1}')
print(f'train_loss : {epoch_train_loss} val_loss : {epoch_val_loss}')
print(f'train_accuracy : {epoch_train_acc*100} val_accuracy : {epoch_
val_acc*100}')
if epoch_val_loss <= valid_loss_min:
    torch.save(model.state_dict(), 'data/temp/state_dict.pt')
    print('Validation loss change ({:.6f} --> {:.6f}).  Saving model
    ...'.format(valid_loss_min,epoch_val_loss))
    valid_loss_min = epoch_val_loss
print('\n')
```

We have added enough print statements to show a clear picture about how much the model learns in each epoch. You will be able to see the logs as follows:

```
Epoch 1
train_loss : 0.6903249303499858 val_loss : 0.6897501349449158
train_accuracy : 54.666666666666664 val_accuracy : 52.400000000000006
Validation loss change (inf --> 0.689750).  Saving model ...

Epoch 2
train_loss : 0.6426109115282694 val_loss : 0.7218503952026367
train_accuracy : 64.8 val_accuracy : 57.199999999999996
```

We can also visualize these in a chart as shown in Figure 16-5.

```
import matplotlib.pyplot as plt

fig = plt.figure(figsize = (20, 6))
plt.subplot(1, 2, 1)
plt.plot(epoch_tr_acc, label='Train Acc')
plt.plot(epoch_vl_acc, label='Validation Acc')
plt.title("Accuracy")
plt.legend()
plt.grid()

plt.subplot(1, 2, 2)
plt.plot(epoch_tr_loss, label='Train loss')
plt.plot(epoch_vl_loss, label='Validation loss')
```

```
plt.title("Loss")
plt.legend()
plt.grid()

plt.show()
```

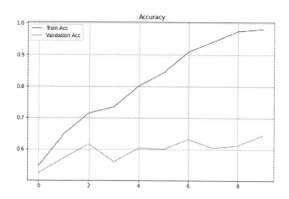 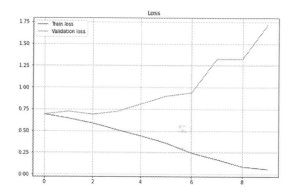

Figure 16-5. *Accuracy and loss over epochs*

For further improvements and tuning, you can play with the model architecture and hyperparameters and, the easiest of all, increase the number of epochs or add more labelled data.

Serializing for Future Predictions

Usually, you'd play around with modifications in terms of network architecture, tune factors like how you are creating features (vocabulary), and other hyperparameters. Once you've got sufficiently high accuracy that can be reliably used in the application, you would save the model state so that we don't have to repeat computationally intensive training process every time we want to predict sentiment for a sentence.

Assuming that we will not use the training method again, here are the items that need to be stored for the prediction phase:

- Logic to convert sentence into sequence

- Vocabulary dictionary containing mapping from words to numbers

- Network architecture and forward propagation computations

Python's pickle module is often used to serialize objects and store them permanently in the disk. PyTorch provides a method to save the model parameters, which internally uses pickle by default, and models, tensors, and dictionaries of all kinds of objects can be saved using this function.

```
torch.save(model.state_dict(), 'model_path.pt')
```

state_dict is a Python dictionary object that maps each layer to its parameter tensor. In future, you can load the parameters using

```
model = SentimentAnalysisModel (args)
model.load_state_dict(torch.load('model_path.pt'))
```

Remember, this only saves the model parameters. You would still need the model definition that you have specified in the code before.

The inference method can use the vocabulary dictionary to convert sequences of word tokens to sequences of numbers. We will pad this up to the length we decided before (200) and use that as input for the model. This method is implemented as follows:

```
def inference(text):
    word_seq = np.array([vocab[preprocess_string(word)] for word in text.
    split() if preprocess_string(word) in vocab.keys()])
    word_seq = np.expand_dims(word_seq,axis=0)
    padded =  torch.from_numpy(pad(word_seq,500))
    inputs = padded.to(device)
    batch_size = 1
    h = model.init_hidden(batch_size)
    h = tuple([each.data for each in h])
    output, h = model(inputs, h)
    return(output.item())
```

You can simply call this method to find the sentiment score. Make sure to call `model.eval()` if you want to make an inference.

```
inference("The plot was deeply engaging and I couldn't move")
```

This will preprocess the sentence, split it into tokens, convert into a sequence of vocabulary indices, pass the input sequence to the network, and return the value obtained in the output layer after a forward propagation pass. This returned a value of 0.6632 , which denotes a positive sentiment. If the use case requires, you can add a conditional statement to return a string containing a word "positive" or "negative" instead of a number.

Hosting the Model

One of the highly popular methods to use a train model in a larger application is to host the model as a microservice. This means a small HTTP server will be used that can accept GET requests.

In this example, we can build and create a server that can accept GET data, which will be a review. The server will read the data and respond with a sentiment label.

Hello World in Flask

If you haven't installed Flask, you can install it using

```
pip install Flask
```

Here's a very simple hello world application in flask. You will need to create an object of Flask() and implement a function that is used by an address defined by @ app.route(). Here, we're defining an endpoint that will accept a request to the server URL which will be represented as http://server-url:5000/hello and returns a string saying Hello World!. We have to specify

```
from flask import Flask
app = Flask(__name__)

@app.route('/hello')
def hello():
    return 'Hello World!'
app.run(port=5000)
```

If you run it on Jupyter or terminal, you will be able to see the server logs:

```
* Serving Flask app "__main__" (lazy loading)
 * Environment: production
   WARNING: This is a development server. Do not use it in a production
   deployment.
   Use a production WSGI server instead.
 * Debug mode: off
```

You can now navigate to `http://127.0.0.1:5000/hello` on the browser and confirm the output. The server logs will confirm that.

```
127.0.0.1 - - [28/Jun/2021 10:10:00] "GET /hello HTTP/1.1" 200
```

To host the model, you can create a separate Python file and define a new

```python
@app.route("/getsentiment", methods=['GET'])
def addParams():
    args = request.args
    text = args['reviewtext']
    score = inference(text)
    label = 'positive' if score >0.5 else 'negative'
    return {'sentimentscore': score, 'sentimentlabel':label}
```

The front-end application can send a request to `http://server:port/getsentiment` and send the data as a reviewtext argument and receive a json/dictionary with sentimentscore and sentimentlabel.

What's Next

The field of machine learning, artificial intelligence, and data science has been evolving over the past decades and will keep evolving as newer hardware technologies and algorithmic perspectives keep evolving.

AI is not a magic wand that will solve our unsolvable problems – but a well-structured suite of concepts, theories, and techniques that help us understand and implement solutions that help the machines learn by looking at the data that we offer them. It is important to understand the implications of potential biases and thoroughly inspect the ethical aspects of the projects and products that are the outcome of our practice. This book serves not as an end but as a handy tool to navigate the steps in your data science journey.

Index

© Ashwin Pajankar and Aditya Joshi 2022
A. Pajankar and A. Joshi, *Hands-on Machine Learning with Python*, https://doi.org/10.1007/978-1-4842-7921-2

Printed in the United States
by Baker & Taylor Publisher Services